Teaching Human Variation: Issues, Trends and Challenges

TEACHING HUMAN VARIATION: ISSUES, TRENDS AND CHALLENGES

GORAN ŠTRKALJ
EDITOR

Nova Science Publishers, Inc.
New York

LIBRARY OF CONGRESS CATALOGING-IN-PUBLICATION DATA
Teaching human variation : issues, trends and challenges / editors, Goran Štrkalj.
p. ; cm.
Includes bibliographical references and index.
ISBN 978-1-60876-616-1 (hardcover)
1. Physical anthropology--Study and teaching. 2. Human genetics--Variation. I. Štrkalj, Goran. [DNLM: 1. Anthropology, Physical--education. 2. Continental Population Groups--classification. 3. Continental Population Groups--genetics. 4. Evolution. 5. Genetic Variation. 6. Prejudice. GN 269 T253 2009]
GN59.6.T43 2009
599.9071--dc22
2009044354

Published by Nova Science Publishers, Inc. ✢ New York

CONTENTS

FOREWORD: RANDOM THOUGHTS ON TEACHING HUMAN VARIATION, PAST AND PRESENT

Phillip V. Tobias*

University of the Witwatersrand, Johannesburg, South Africa

It is fitting that this book on *Teaching Human Variation* was planned for 2009, for that year marked the bicentenary of the birth of Charles Darwin and the sesquicentenary of the publication of *The Origin of Species by Means of Natural Selection.* The concept of Natural Selection was put forward to explain how evolutionary change might occur. We owe this hypothesis to Charles Darwin and Alfred Russel Wallace. Independently, they lighted upon it and presented their preliminary essays to the Linnean Society in London on 1st July 1858.

All teaching of biological variation should start by reference to evolution and what Wallace, in an act of extraordinary generosity, proposed should be called Darwinism.

There is a tendency by some scholars to inflate the list of characters to lay at the door of the race concept: We caution them against taking spurious comfort from, and reading too much into, the full title of Darwin's 1859 book: *The Origin of Species by Means of Natural Selection or the Preservation of Favoured Races in the Struggle for Life.* Such comfort is 'spurious' for two rea

* Professor Emeritus Phillip V. Tobias* F.R.S., Hon.M.NAS, Hon. F.R.S.S.Af. University of the Witwatersrand, Johannesburg, South Africa. E-mail: Phillip.Tobias@wits.ac.za

sons. Darwin's use of the term 'favoured races' is not a statement of value judgment (as in, "the brain of Race A is better developed than that of Race B"). No, far from such thinking, Darwin's use of the term 'favoured races' refers to the idea that some populations are more favoured by Natural Selection, that is, they may have a greater chance of survival and of reproduction than less favoured races.

The second reason it would be gratuitous for racists to be encouraged by the title of Darwin's 1859 book is that the volume does not discuss the human species – save for a single short sentence on the last page[1]. So the book bearing what some may consider a controversial title does not deal with human variation and zealously eschews discussion of human evolution. Only after another twelve years elapsed did Darwin address those themes in *The Descent of Man and Selection in Relation to Sex* (1871).

Goran Štrkalj's book was planned for just 150 years after *The Origin of Species* first saw the light of day. I began this Foreword by emphasising this interval, for no comprehensive teaching of human variation can occur without it being considered as a function of *time*. Deep time is the domain of evolutionary change, or phylogeny, the direct evidence for which is mainly the palaeontological record in the rocks that make up the earth's crust. Recent time refers to more recent archaeological and fossil remains and to living peoples and their ontogeny. Human variation represents a history of over 5 million years and, on the latest calculations, perhaps 10 million years.

Štrkalj's book brings together carefully-analysed and -formulated chapters by a number of authors. These chapters address many aspects of the study of human variation. Collectively, they show that variability is a fundamental property of living things and the fossil evidence indicates that it occurs also in ancient life-forms. Some of the chapters address the intrusion in discussions of natural variation of value judgments, especially in the study of humans, hominins and hominids. Indeed, it is not cynical to point out that the closer the objects of study are to human species past and present, the more frequently have evaluative or judgmental factors been imported into the assumptions and causal analyses made about such variability. In the twentieth century, especially, these value judgments were accompanied by racial classification and oftentimes transmuted into aspects of racialism and racism, as well as into tragic political policies.

An important task for teachers of human variation is to aspire to become amphibious. I borrow this term from Joseph Needham's volume of essays and

[1] "Much light will be thrown on the origin of man and his history."

addresses, *The Great Amphibium* (1931). Teachers of human variation should swim in the ocean of human biology while walking on the shorelines of evolution. Thus, I favor the development of linkages between biological concepts that refer to living human beings and the possible evolutionary bases. This is partly why I threw myself wholeheartedly into organising the so-called Dual Congress a few years ago. Both the International Association of Human Biologists and the International Association of Human Palaeontologists had approached me regarding meetings. Instead of agreeing to organise two meetings, I proposed a Dual Congress, devoted to discussion of both the living and the dead. We hoped thereby to bring together in the various symposia scholars of current biological issues – such as morphology, demography and ecology, as well as of the past, as in palaeomorphology, palaeodemography and palaeo-ecology.

In preparing this Foreword, it struck me that consideration of secular trends in human growth and development might be a valuable means by which to approach the two faces of Eve – and of Adam – along the pathway of variational studies.

I propose the use of the term *secularity* to connote the propensity of a population to show secular biological changes of one or more characteristics.

The Meaning of the Term Secular Trend

It is not, I am sure, necessary to remind you that the word 'secular' is derived from the Latin adjective '*saecularis*'; this in turn comes from the Latin noun, '*saeculum*', meaning an age, a generation, a century, a very long time. The Secular Games were an Ancient Roman festival held every 100 or 120 years. Hence, the term 'secular' indicates duration for an age or for a hundred years, as distinct from short-term or periodical. Thus, a secular change is understood as a change that is evident over a long period: long in relation to the length of a generation; but it does not imply long in geological time.

In Human Biology the term 'secular trend' has been applied mainly to the marked trend during the last 125 years for the physical maturation of children in many, especially First World populations, to occur more rapidly. The trend manifests itself essentially in two ways: first, in stature and mass, children have become appreciably larger at each age period than they formerly were at the same age. Secondly, the phenomena of puberty, especially the adolescent growth-spurt and menarche, occur at progressively younger ages. Most studies on the secular trend in *Homo sapiens* are based on statistical data for the age at menarche and for growth in stature and mass.

However, along the way, the term secular trend acquired a wider connotation. Today, it applies to long-term systematic or non-random changes in a wide variety of traits, in successive generations of a population living in the same location (as defined by Napoleon Wolanski of Poland in 1966 and 1967). The emphasis on *the same place of residence* excludes from the secular trend changes attributable to the effects of migration from one area to another. Such migration effects include changes that may accompany urbanisation of previously rural members of a population or community.

CHARACTERISTICS THAT SHOW A SECULAR TREND

Some twenty-two characteristics are found to show secular changes in some human populations. These include 6 traits of newborn babies, 7 traits of immature individuals during the growing period (from the neonatal to the young adult stage), and 9 characteristics of adult individuals (Tobias, 1978). I found that it is misleading to speak simply of the secular trend. I clarified the concept of *direction* in secular change and identified instances of *positive secular trends*, *negative secular trends* and *absent secular trends* (Tobias & Netscher, 1976, 1977; Tobias, 1985).

THE CAUSES OF THE SECULAR TREND

James Tanner (1962) summarises various causal factors proposed for the positive secular trend towards faster growth and bigger bodily size:

1. Nutritional improvement – to which we might add, a better balance between diet and exercise.
2. General improvement in environmental circumstances, that is, a broader range of environmental factors than diet alone.
3. The gradual rise in world temperature since the 1850s and, particularly, since 1920 (Mills, 1937, 1949).
4. A possible genetic cause operating in populations where reproductive isolates have been breaking down, with an increase in outbreeding and

resultant heterosis or hybrid vigour (Broman *et al.*, 1942, supported by the researches of Hulse, 1957, and Lasker, 1960).

Factors (1) and (2) find the most widespread support among human biologists today as the principal causes of the (positive) secular trend. Moreover, a body of evidence indicates that a genetic factor may operate in some populations, perhaps in addition to environmental factors. The world temperature effect is controversial: for example, Tanner (1962) points out that all evidence controverts the idea that a hot climate accelerates growth and that Mills's prediction of a reversal of the secular trend of accelerated growth in the 1940s and 1950s was not realised. Another strong argument against the supposed temperature effect is the fact that, even while the world temperatures have been rising, some human populations show a positive secular trend, while others show an absence of this secular trend and still others, show a negative secular trend (i.e., slowed growth).

A problem that interests me is this: does the same set of causal factors operate on all the characteristics that may change in a secular fashion? That is, are the same causal factors responsible for some 22 traits shown to change in a secular manner? for example, for the acceleration of growth processes in a population and for the increase of mean adult body size? If a population shows a trend towards larger neonatal birth-weight, does it follow that the same population would show more rapid maturation and larger adult bodily size? Are these various traits indissolubly correlated biologically, or are some traits in our list of 22 characteristics dissociable from others?

As long ago as 1942, Broman *et al.* suggested that the causal factors responsible for a positive secular trend during the growth phase were indeed different from those operating upon the positive secular trend of adult size. They opine that acceleration of the growth process represents a response to environmental agents, while the increase in adult size is genetic in origin and the result of a breakdown in marital isolates, with consequent outbreeding and heterosis. However, an increase in adult size is found also under conditions where there seems to be little evidence of a dramatic increase in outbreeding, such as among the South African San or Bushmen (Tobias, 1962) and Aboriginal Australian communities (Barrett & Brown, 1971). Thus, populations may differ from one another in the concurrence of causal factors that operate to produce secular changes.

Of interest is the particular case of deteriorating conditions of life during wartime. This deterioration illustrates that secularity and environmental impoverishment are interdependent.

WAR AND THE SECULAR TREND IN MODERN HUMAN POPULATIONS

Data on South African black populations show a negative secular trend over the past 100 years. In re-examining data from two studies on the length of femora from cadavera dissected in the Wits Anatomy Department (Tobias & Netscher, 1977; Price *et al.*, 1986) and from another investigation on age-ranked stature data of Johannesburg black people (Tobias, 1986), I found three distinct phases of this negative secular trend, each seeming to follow a phase of physical melioration. The first phase of decline coincides with the Anglo-Boer War of 1899-1902, with subsequent improvement during a period of reconstruction leading to the First World War. A second phase of negative secular trend appears in the years 1919-1920, following the end of the First World War. Our results for the third period are somewhat conflicting. Data from living Johannesburg black people from the period between the two world wars, suggests melioration, whereas data from the cadavers suggest a further decline from 1922 to 1932. The third period covers the years 1939-1949, i.e., the decade encompassing the Second World War (Tobias, 1988).

In short, the three phases demonstrate a decline in adult mean stature. All three phases coincide with three major wars affecting *inter alia* the human populations of southern Africa. These synchronous phenomena do not exclude other socio-economic factors as possible concomitant causes of the detected physical decline.

Earlier I provided what may well be the first evidence that a cadaveric population is a *more sensitive indicator* of adverse circumstances than a random cross-section of the same living population. The cause is likely a simple one: adverse conditions may be assumed to affect deprived strata of society (from which most cadavera emanate) more severely than more affluent strata. The affluent sections of a population, it would seem, are protected by their very affluence from too blatant exposure to conditions predisposing to negative secular effects. In short, we hypothesise that in a population showing a negative secular trend, the poor suffer negative secular effects more markedly than those who are well-off!

This observation may explain the fact that wartime and post-war conditions seem to generate smaller negative secular trends in a cross-section of the population and larger negative secular trends in a non-random, pauperised sector, such as that represented by the unclaimed bodies forming the cadaveric population.

This finding is the converse of the already known varying intensity of *positive* secular effects in varying social strata. Referring to *the* secular trend—by which, it is clear he means a *positive* secular trend, Miroslav Prokopec (1984: 58-59) sums up his position:

> This trend operates with dissimilar intensity in different countries during various times, and it may differ between the privileged and underprivileged classes of the same industrialised population.

SECULARISM AND EVOLUTION: A NEW HYPOTHESIS

With Prokopec, let me raise the possibility that the nexus of causal factors that operate to produce secular effects may differ from one population to another. I now advance the hypothesis that as populations may vary in their degree of secularity, such variances may be of evolutionary significance.

Various human populations or segments of populations respond to adverse environmental circumstances to varying degrees. The negative secular trend has varied in intensity between an average, cross-sectional series and a putatively lower social stratum of the South African black population. An analysis of such differences is difficult and nuanced because of other differences among recent human populations – such as educational level, socio-economic status, political repression as under slavery or apartheid, dietary regimen and customs, and other environmental and cultural factors. If one were able to equalise or correct for all such factors, it remains possible that populations may differ when exposed to the same patterns of adversity. That is, perhaps there are differences in secularity, or in the degree to which different populations respond to the same set of environmental challenges.

If such differences in secularity could be determined, a new theoretical approach to the analysis of secularity might be proposed. For instance, population A may show severe human biological effects in response to environmental deterioration. That is, the sensitivity of such a population might be much greater than that of other populations. Population B may manifest a degree of insensitivity or resistance to the same set of impoverishments. We may tentatively identify these two hypothetical populations as *secularity-sensitive* and *secularity-resistant* respectively.

If data demonstrate such variances in living human populations, no matter what the cause of such variability, it is likely that the two populations are imbued with varying degrees of flexibility or stability in the face of long-term environmental changes. In other words, secularism and varying secularity may intervene as other "layers" between the environment and the human organism.

Similarly, the wide range of functional modalities provided by the human brain may be regarded as a "layer", making for greater flexibility and adaptability in the evolution of the hominid clade. By analogy, it would be valuable to know whether there is a trend towards an increase or a decrease in mean secularity in the emergence and micro-evolution of humanity. Selection likely favours more flexible patterns of survival and adaptability over less flexible and less adaptable populations of evolving humanity. As a theoretical construct, therefore, it might be useful for researchers and teachers to address the evolutionary implications of secularity. Secularity may widen the range of ecological niches within which hominid survival and adaptation might occur. In that case we might see a role for the natural selection of a diverse array of secularities among evolving and diversifying humankind. I say "as a theoretical construct" because, to my knowledge, we do not yet possess the analytical data or tools with which to convert this idea into a set of testable hypotheses.

For teachers of human variation, these few ideas may help to enliven teaching sessions and projects. I hope so! We should all be grateful to Goran Štrkalj, and the dozen or so other contributors, for creating this book. I wish him and the other authors great success. The imparting of any knowledge is no less crucial than the advancement of its frontiers. Those who – amphibiously – can succeed in both impulses will, I am confident, have the makings of teachers of distinction.

References

Barrett, M.J., & Brown, T. (1971). Increase in average height of Australian Aborigines. *Medical Journal of Australia* 2: 1169-1172.

Broman, B., Dahlberg, G., & Lichtenstein A. (1942) Height and weight during growth. *Acta Paediatrica* 30: 1-66.

Darwin, C. (1859). *The Origin of Species by Means of Natural Selection.* London: John Murray.

Darwin, C. (1871). *The Descent of Man and Selection in Relation to Sex.* London: John Murray.

Hulse, F.S. (1957). Exogamie et hétérosis, *Archives Suisses d'Anthropologie Générale:* 22: 103-125.

Lasker, G.W. (1960). Variances of bodily measurements in the offspring of natives and of immigrants to three Peruvian towns. *American Journal of Physical Anthropology,* n.s. 18: 257-261.

Mills, C.A. (1937). Geographic and time variations in body growth and age at menarche. *Human Biology* 9: 43-56.

Mills, C.A. (1949). Temperature dominance over human life. *Science* 110: 267-271.

Needham, J. (1931). *The Great Amphibium.* London: Students' Christian Movement Press.

Price, B., Cameron, N., & Tobias, P.V. (1987). A further search for a secular trend of adult body size in South African Blacks: evidence from the femur and tibia. *Human Biology* 59: 467-475.

Prokopec, M. (1984). Secular trends in body size and proportions, and their meaning. *Studies in Human Ecology* 6: 37-61.

Tanner, J. (1962). *Growth at Adolescence* (2nd edition). Oxford: Blackwell Scientific Publications.

Tobias, P.V. (1962). On the increasing stature of the Bushmen. *Anthropos*: 57: 801-810.

Tobias, P.V. (1970). Puberty, growth, malnutrition and the weaker sex – and two new measures of environmental betterment. *The Leech* 40: 101-107.

Tobias, P.V. (1975). Stature and secular trend among Southern African Negroes and San (Bushmen). *South African Journal of Medical Sciences* 40: 145-164.

Tobias, P.V. (1975). Anthropometry among disadvantaged peoples: studies in Southern Africa. In E.S. Watts, F.E. Johnston, & G.W. Lasker (Eds.), *Biosocial Interrelations in Population Adaptation* (pp. 287-305). The Hague: Mouton.

Tobias, P.V. (1978). The secular trend in African peoples. In A.J. Brink & C.M. Lewis (Eds.), *Mankind and Medicine in the Third Millenium* (pp. 64-74 and Discussion pp. 92-93). Parrowvalei: University of Stellenbosch.

Tobias, P.V. (1985). The negative secular trend. Special issue in honour of Professor J. Huizinga on his retirement. *Journal of Human Evolution* 14: 347-356.

Tobias, P.V. (1986). Physical stature in disadvantaged communities: Johannesburg blacks have not grown taller this century. *South African Journal of Science* 82: 585-588.

Tobias, P.V. (1988). War and the negative secular trend of South African blacks with observations on the relative sensitivity of cadaveric and non-cadaveric populations to secular effects. In *Proceedings of the 5th*

Congress of the European Anthropological Association, 1986 (pp. 451-461). Lisbon: EAA.

Tobias, P.V., & Netscher, D. (1976). Evidence from African Negro skeletons for a reversal of the usual secular trend. *Journal of Anatomy* 121: 435-436.

Tobias, P.V., & Netscher, D. (1977). Reversal of the usual secular trend, as evidenced by South African Negro crania and femora. *Mitteilungen der Anthropologischen Gesellschaft (Wien)* 103: 233-239.

Wolanski, N. (1966). The secular trend; microevolution, physiological adaptation and migration, and their causative factors. In *Problems of World Nutrition* (pp. 96-108). Proceedings of the 7th International Congress of Nutrition, Hamburg, 1966.

Wolanski, N. (1967). Basic problems in physical development in man in relation to the evaluation of development of children and youth. *Current Anthropology* 8: 35- 60.

INTRODUCTION

Goran Štrkalj[*]

Department of Chiropractic, Macquarie University, Sydney, Australia
Institute for Human Evolution, University of the Witwatersrand, Johannesburg, South Africa

Human biological variation has been one of the main research subjects of biological (physical) anthropology since its inception as a distinct scientific discipline. While humans vary biologically in different ways and at various levels of biological organisation (e.g., individually, due to sex, age, etc.), it appears that in anthropology, human variation due to ancestry is most vigorously investigated and has provoked the most intense debates and controversies. For a lengthy period, the concept of race was the key for describing and understanding this mode of human variation. Indeed, it is the concept of race that has provoked most of the controversies throughout history. Numerous issues concerning race were debated, including the criteria for classification, the number and possible hierarchical nature of human races, their origins and their relationships to each other. Some of these debates, such as the polemic on the most accurate classification of humanity into races, lasted for decades in the same or slightly modified form. Charles Darwin (1871: 270), for example, in his *Descent of Man* notes that "Man has been studied more carefully than any other animal, and yet there is the greatest possible diversity among the capable judges" on how the classification was carried out, with the number of races varying from one to sixty-three. Many

[*] E-mail: goran.strkalj@mq.edu.au

anthropologists, such as the Russian-born French scientist Joseph Deniker, author of the classical work *The Races of Man*, saw no problem in these disagreements but perceived them rather as proof of the growth of scientific knowledge. In Deniker's (1900: 280) words: "In all sciences, classifications change in proportion as the facts or objects to be classed become better known". The process and the great amplitude of disagreement, however, continue far beyond Deniker's expectations and, almost a century after Darwin, Ernst Mayr (1970: 395) makes a comment that is almost identical to Darwin's: "There is no agreement yet on the formal classification of the subdivisions of *Homo sapiens*. Of two highly competent recent treatises on the races of man, one recognized 6 races, the other recognized 30."

The debate on race acquires a new dimension in the mid-twentieth century when the scientific validity of the concept is questioned and alternative approaches to the study of human variation begin to appear. At least two groups of anthropologists interested in human variation emerged in the late 1950s: those who accepted the concept of race and those who rejected it in favour of new approaches such as clinal. Since then, most anthropologists accept the non-racial position, although a significant minority (in some countries a majority) of experts supporting the concept of race remains (Kaszycka *et al.*, 2009). Research shows many diverse reasons for this and other disagreements in the field of human variation studies. One reason is educational background. It appears that scientific attitudes towards human variation and race vary significantly according to the research and teaching tradition to which a particular scientist belong to (Kaszycka *et al.*, 2009). This tendency becomes even more significant as discussions on subjects related to human variation transcend the boundaries of biological anthropology and enter other disciplines, such as the biomedical sciences (Štrkalj *et al.*, 2004). Indeed, it is suggested that some of the misunderstanding within biological sciences concerning the highly controversial subject of race is due to a lack of knowledge of the nature of human variation on the part of scientists (Štrkalj & Wilkinson, 2006).

Bearing all this in mind, we may argue that the teaching of human variation to students of biological anthropology and other disciplines is of considerable importance. It appears, however, that the literature on teaching about human variation is relatively meagre in both scientific and educational sectors. In this book, the authors, who are biological anthropologists and educators, investigate different aspects of teaching human variation. The topics, approaches, scopes and focus of the contributed papers vary considerably. They include first-hand experience of leading pedagogues, empirical studies establishing efficacy of teaching, recommendations on how to improve teaching, as well as reflections on the place of the study of human variation within broader scientific and social contexts. It is hoped that the book, thus structured, will contribute to the improvement of teaching human variation by raising issues that might be of interest to a wide range of experts involved in education.

Phillip Tobias shows in his Foreword how secular trends in human growth and development may be used in understanding human variation, and introduces a new term "secularity" to connote "the propensity of a population to show secular biological changes of one or more characters". Chapter One starts with Darren Curnoe's Dobzhanskyan statement that "Nothing in biological anthropology makes sense in the absence of a proper understanding of human variation". This programmatic assertion underpins all chapters in the present volume. Curnoe also explains why human variation is the central issue of biological anthropology and goes on to show the significance of this subject to other scientific disciplines. One sub-discipline of biological anthropology where human variation is of paramount importance is forensic anthropology. This discipline, which has gained considerable media attention and great interest from the general public, focuses on determination of race as one of its key tasks. The flaws of this approach are investigated by Alan Morris in a chapter concentrating on the ever complex context of South Africa. Maciej Henneberg, drawing on his experience teaching biological anthropology on four continents, highlights the failure of anthropologists to change public perception on human variation and the inefficiency of teaching human variation in tertiary institutions. He argues for the inclusion of this subject in the curricula of primary and secondary schools and for more proactive engagement of anthropologists in educating the general public through the media and other means. Rachel Caspari examines the complex relationship between race and geographical variation and argues that conflating the two might impede the teaching. Patricia Rice presents an in-class exercise for teaching human variation, tackling some of the core concepts and issues related to the subject. Pamela Ashmore and Donna Hart focus on inquiry-based courses on human variation and their outcomes. They demonstrate how significant changes in student knowledge and perceptions could be achieved. Research in human variation has a long and turbulent history, whose importance for students of anthropology is argued by a number of authors. Goran Štrkalj and Robyn Beirman propose a way in which history of science could be incorporated in teaching human variation. Several chapters argue that the relevance and implication of human variation transcend the boundaries of biological anthropology and its subdisciplines. One of the fields where human variation seem to be of intense interest is medicine. Tracey Wilkinson, Goran Štrkalj and Muhammad Spocter argue that human variation should be taught to medical students. The relevance of human variation studies, however, transcends the fuzzy borders of science: The social implications of research on human biological variations are numerous and far-reaching. Charles Susanne focuses on some of these implications in his analysis of various forms of racism and ways to counteract them.

Educating students (and eventually general public) about the human biological variation continues to be one of the greatest academic challenges. The essays presented in this book provide responses to this challenge and suggest ways to improve teaching and facilitate a better understanding of human biological variation.

REFERENCES

Darwin, C.R. (1871). *The Descent of Man and Selection in Relation to Sex*. London: John Murray.

Deniker, J. (1900). *The Races of Man: An Outline of Anthropology and Ethnography*. London: Walter Scott.

Kaszycka, K.A., Štrkalj, G., & Strzałko, J. (2009). Current views of European anthropologists on race: influence of educational and ideological background. *American Anthropologist* 111: 43-56.

Mayr, E. (1970). *Populations, Species and Evolution*. Cambridge, Mass.: Harvard University Press.

Štrkalj, G., Ramsey, S., & Wilkinson, A.T. (2004). Anatomists' attitudes towards the concept of race. *South African Medical Journal* 94: 90-91.

Štrkalj, G., & Wilkinson, A.T. (2006). "Race" and biomedical research: an educational perspective. *Education for Health* 19(1): 111-114.

In: Teaching Human Variation
Editor: Goran Štrkalj
ISBN: 978-1-60876-616-1

Chapter 1

HUMAN VARIATION: THE MAJOR UNIFYING THEME OF BIOLOGICAL ANTHROPOLOGY

Darren Curnoe *

School of Biological, Earth and Environmental Sciences,
University of New South Wales, Sydney, Australia

ABSTRACT

Nothing in biological anthropology makes sense in the absence of a proper understanding of human variation. The study of variation has been and remains, therefore, the core, unifying theme of this scientific discipline. Moreover, it provides a fundamental link with the theory of evolution; given the central role that variation plays in our understanding of evolution and the core place of this theory in modern biological science. This chapter provides a brief overview of some major historical aspects of 'race' research in biological anthropology. It also examines the importance of human variation to contemporary research in biological anthropology, providing an example provided from palaeoanthropology. I also explore briefly the role of variation as a basis for equipping (pedagogically) students for an increasingly inter-connected, multicultural and ethically engaged community, a planet facing major challenges such as anthropogenic climate change, and emerging fields in the morphological and molecular sciences. Finally, I propose that the study of variation and diversity provides opportunities for educators to meet the new phase of university teaching with its emphasis on generic attributes.

INTRODUCTION

Understanding how and why humans vary genetically and phenotypically (i.e. physically and behaviourally) continues to be of fundamental importance to the scientific field of biological (physical) anthropology. Paradigms and methods of approaching these questions have changed dramatically over the last 50 years or so – especially since the biological 'modern synthesis' began to influence anthropology from about the 1950s onwards. Today, the question of variation holds a key place in terms of understanding the history of the discipline as well as underpinning all major contemporary fields of research in biological anthropology.

The teaching of human variation is both rewarding and challenging. After a long period of introspection – almost embarrassment – about a historical preoccupation with classifying humans into racial groups, anthropology is now well placed to meet the educational challenges of the 21st century. The knowledge and skills base of biological anthropology are increasingly relevant to students in a world facing global challenges such as climate change. Research in anthropology is also at the leading edge of medical, morphological, evolutionary and molecular sciences. The discipline also provides unique opportunities to meet the needs of students facing rapidly changing, increasingly diverse, ever more interconnected, and more ethically oriented workplaces and communities.

This chapter reflects a mixture of the author's experiences, observations and reflections over more than a decade of research and teaching in biological anthropology and related fields including anatomy, palaeontology, geology, genetics and archaeology. This material also stems from my experience working in two countries at three universities – each with highly diverse academic and student bodies, and from many years overseas pursuing research activities.

This chapter aims mainly to demonstrate both the historical importance and the increasing relevance of the teaching and research of human variation. Examples are provided to support this argument and to illustrate ways in

* E-mail: d.curnoe@unsw.edu.au

which teaching variation might meet the vocational needs of and challenges facing students in 21st century universities, workplaces and communities.

SOME HISTORICAL THEMES

A great deal has been written about the history of race in both scientific and popular literature: the short review presented here barely skims the surface of a complicated and demanding subject and aims only to touch on a few key themes. Two provocative and well-written popular science books include Gould's (1981) *The Mismeasure of Man* and Kohn's (1996) *The Race Gallery.* In addition, a substantial body of critical scientific literature is accessible in popular science magazines such as *Scientific American* and *New Scientist,* and in academic journals *Current Anthropology*, *American Journal of Physical Anthropology, Evolutionary Anthropology* and *Nature Genetics,* among others.

During the latter 19th and early-mid 20th centuries, racial typology – the classification of humans into racial groups based on an ideal form or 'type' – grew out of the colonial experience. It was to become a preoccupation of physical anthropologists and grew(e.g. Hooton, 1926) until the mid-20th Century and provided both the theoretical foundation, methodologies and lexicon for the study of human variation. Yet, it seems that little of the "science of race" was founded in the scientific method, as we understand it today:, a fact frequently unappreciated by even professional anthropologists. Rather, the study of race at that time seems to reflect a mixture of categorisation based mostly on spoken language (i.e., linguistic affiliation), observed cultural differences and a few general, supporting observations of the physical characteristics of particular groups. There was little regard if any for variability within groups, which, when observed, was regarded by researchers as representing admixture with other races, rather than geographical variation or variation among individuals owing to differences in lifestyle, diet or childhood disease burden.

For example, in his introduction to *Races of Africa* (1959: 1), Seligman remarks:

> It is obvious that questions of race should first and last be determined by the study of physical characters, yet in no part of Africa is there in existence anything approaching an anthropological survey (based on such characters),

> nor can it be said for any considerable area that even the first rough survey work has been done.

Despite this acknowledged paucity of data, three editions with a reprint of the third edition of Seligman's book were published between 1930 and 1959 (Note: the quote above is from the 1959 edition, only 50 years ago). Put simply, in those years, to speak of the "science of race" was to utter an oxymoron.

For many people, the study of human variation continues to be synonymous with the categorisation of people into races. Yet, the study of variation over the last 50 years or so has moved well beyond mere pigeonholing of variation into geographic groups to encompass an understanding of how and why humans, from individuals to populations, vary across space and through time and at both phenotypic and molecular levels.

The application of the category of race to *Homo sapiens* was strongly questioned and firmly rejected by many anthropologists in the late 20th century. During the 1980s and 1990s, physical anthropologists began to abandon the concept in droves. Surveys reveal that half of those surveyed rejected the existence of human races, while most U.S. college textbooks rejected the concept entirely (Lieberman & Jackson, 1995). In 1996, for example, the American Association of Physical Anthropologists, the largest professional association for physical/biological anthropologists in the world, published a position statement entitled the *Biological Aspects of Race* (AAPA, 1996). A couple of quotes illustrate the subject and tenor of this document:

> 3. There is great genetic diversity within all human populations. Pure races, in the sense of genetically homogenous populations, do not exist in the human species today, nor is there any evidence that they have ever existed in the past.
>
> 5... Generally, the traits used to characterize a population are either independently inherited or show only varying degrees of association with one another within each population. Therefore, the combination of these traits in an individual very commonly deviates from the average combination in the population. This fact renders untenable the idea of discrete races made up chiefly of typical representatives.

Much of the disquiet surrounding the classification of humans into races likely originates from the confluence of three factors during the mid-20th century: (1) the explicit abuses of anthropology by the Nazis and others before and during World War II, (2) the many and major human rights abuses and crimes committed during the 20th century and the associated establishment of the United Nations, its agencies like UNESCO and the production of documents like the *Universal Declaration of Human Rights*, and (3) the modern 'biological synthesis' theories attributed to such 20th century greats as Huxley, Dobzhansky, Wright, Mayr and Simpson which placed the study of variation at the core of evolutionary biology (including anthropology) and synthesised genetics and evolution to provide mechanisms for inheritance and change.

Also greatly influential in the 20th century was Lewontin's (1972) article in *Evolutionary Biology* on the apportionment of human diversity. This article demonstrated, as has much subsequent research, that the vast majority of molecular variation (~85%) in humans is found between two randomly selected individuals within a single population; only a small proportion of molecular variation distinguishing major geographic populations (~10%). This finding marked the beginning of the end of the race concept for many anthropologists.

Finally, it was shown that many of the phenotypic characters historically employed to describe races, such as skin pigmentation, head shape and facial form are characterised by considerable evolutionary plasticity, and vary clinally with environmental gradients such as levels of ultraviolet radiation or ambient air temperature. Relethford (2002), for example, found that around 88% of total variation in skin pigmentation is found among major regions, the opposite pattern to most molecular markers (see above). This finding powerfully demonstrates that phenotypic attributes such as skin colour are strongly shaped by the effects of recent natural selection, making them unreliable markers for reconstructing population history and classification (including placing people into races).

This overview of a few recent historical themes of race research demonstrates its overwhelming historical importance to the discipline. It laid the foundation for later 20th and 21st century research programs and to a great extent provided the foundation for much of anthropology as practiced today. Thus, it is clear that the history of biological anthropology should not be hidden or forgotten, but taught and discussed openly, frankly and critically. In fact, this history provides a wonderful opportunity for students to reflect on questions surrounding the values, ethics and roles of science within the broader community and how science shapes, and is shaped by, popular ideas, movements and trends. It also provides an example of how science is ultimately self-correcting, unlike most others form of knowledge, a key reason for its success.

CONTEMPORARY IMPORTANCE OF HUMAN VARIATION

Human genotypic and phenotypic variation - encompassing individual, sexual, spatial, temporal and pathological forms - have become an increasingly important component of anthropology. In effect, understanding of the causes of human variation remains a unifying theme of the discipline. Table 1 summarizes major sub-disciplines of biological anthropology and lists some major research topics for each that are underpinned by a proper understanding of human variation.

Table 1. Importance of human variation to various sub-disciplines of biological anthropology

Sub-Discipline	Some key areas underpinned by human variation
Palaeoanthropology	Classification Adaptation and functional morphology Microevolution and macroevolution Evo-devo Skeletal biology (incl. ageing, sexing etc.)
Human biology	Adaptation and adaptability Growth and development Disease Diet and nutrition Reproduction Nature versus nurture Skeletal biology (incl. ageing, sexing etc.)
Forensic anthropology	Identification of human remains/skeletal biology (incl. ageing, sexing, 'race' etc.)
Anthropological genetics	Population history and population genomics Evo-devo and functional genomics Diseases and disease gene mapping Evolutionary medicine

Example from Palaeoanthropology

Most major areas of research in palaeoanthropology are underpinned by an understanding of human variation. For example, the classification of fossil hominins at the species level quite rightly remains an important topic of research; as without a classificatory scheme, it is impossible to hypothesise about phylogeny and evolutionary pattern and process, including the origins of modern humans. The practice of classification is itself underpinned by an understanding of normal variation within species (individual, spatial and temporal).

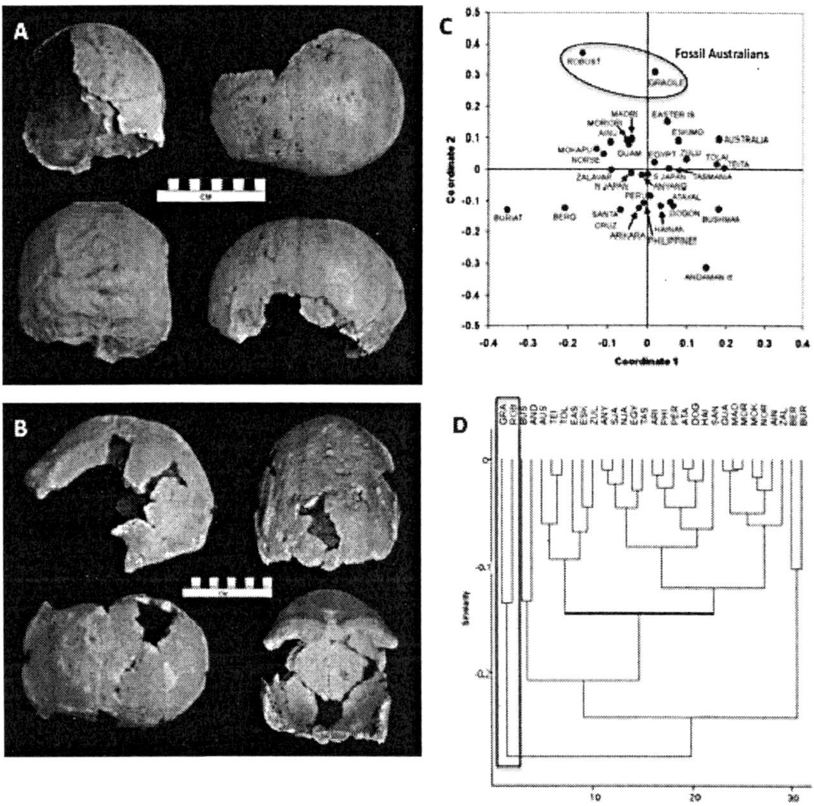

Figure 1. Pleistocene Australian human calvarii: Willandra Lakes Human 3 (A) and 'robust' Willandra Lakes Human 50 (B). Statistical comparisons using canonical variates analysis (C) and cluster analysis (D) demonstrate that the metrical differences between them are larger than those between the most distantly related contemporary human populations. (NB Australian fossil remains shown as: GRA = 'gracile' and ROB = 'robust robust': see text for explanation).

Taxonomists of early hominins such as the australopithecines draw upon yardsticks of intraspecific variation from the non-human apes (i.e., chimpanzees and gorillas). However, the classification of later hominins belonging to *Homo* can be informed and properly executed only by a rigorous understanding of variation within our own species (*H. sapiens*); although, when palaeontological samples eventually become large enough, they will provide crucial supplementary data for classifying fossils, as is the case with *Paranthropus boisei* for early hominins.

Figure 1 provides an example from Australia of the importance of human variation to palaeoanthropology, in this instance, in the reconstruction of the Pleistocene origins of Indigenous Australians. The Willandra Lakes Human (WLH) 3 ('Mungo Man': Fig. 1A) skeleton represents the earliest human remains from Australia, dated to between about 45,000 and 62,000 years old (the dating remains controversial: see Thorne *et al.*, 1999; Bowler *et al.*, 2003). This fossil has been described as possessing a 'gracile' cranial morphology (i.e. lacking a brow ridge, but exhibiting a steep forehead, thin cranial bones, weak muscle markings and small teeth) and as possibly representing colonisation of Australia by an early modern population, possibly from China (Thorne, 1980; Curnoe & Thorne, 2006). In contrast, WLH 50 (Fig. 1B) is dated to around 14,000 years old (Simpson & Grün, 1998), possesses a 'robust' cranial morphology (i.e. possessing a prominent and thick brown ridge, flat forehead, thick cranial bones, strong muscle markings and large teeth) (Thorne, 1980; Curnoe & Thorne, 2006) and is suggested to represent a much later migration to Australia of a population from Southeast Asia (Thorne, 1980). This Pleistocene "dual migration" model is based on the idea that morphological variation between individuals such as WLH 3 and WLH 50 is sufficient to warrant their assignment to different populations (?'races'), each with a different population history, geographic origin and age of colonisation of Australia. Ultimately, contemporary Indigenous Australians are seen as having evolved through a process of admixture between these groups.

How might we assess whether the morphological evidence actually supports such a hypothesis? One way is by undertaking a comparison of variation between these fossil samples (or hypothetical groups) with variation among recent human populations. Figure 1 (C & D) shows the results of descriptive statistical (multivariate) comparisons of the 'gracile' and 'robust' samples (using medians from small palaeontological samples, and Howell's worldwide dataset; see Curnoe & Thorne, 2006). It is clear from both analyses that metrical variation within the Australian human fossil record is very large compared to that seen among recent and geographically disparate humans. So, on face value, and employing a phenetic approach, there seems to be a case for considering the earlier fossils to represent distinct groups of people: the differences between them are larger than that found among even the most geographically distinct populations of recent humans.

An additional aspect of the dual origins model is the claim that the 'robust' Pleistocene group might be descended (at least partially) from hominins who inhabited Southeast Asia from the Lower Pleistocene (classified by most workers in *H. erectus*). Such a scenario is predicted by the Multiregional and Assimilation models (Frayer *et al.*, 1993; Curnoe, 2007). One researcher has even proposed that the presence of the 'robust' cranial form in Australia is evidence for the colonisation of the continent by *H. erectus* itself (Webb, 2005). Although, there is at present no direct evidence for this species having inhabited the region east of Flores in Indonesia.

In order for this theory to hold water, the highly varied morphology of these fossils would need to be shown to be: 1) phenotypically normal (i.e., non-pathological - esp. WLH 50), 2) truly representative of the groups they sample rather than just outliers (difficult to do with very small samples!), and 3) truly the result of different population histories outside Australia rather than adaptation or random genetic drift within a subdivided continental population inhabiting different regions or environments for long periods of time, and in isolation from each other. How might we do this? Again, we could do this by reference to research about the forms and causes of variation among recent/contemporary humans and applying it the Australian Pleistocene record.

In a recent study, the present author (Curnoe, 2009) considered whether the processes of normal growth-development and functional stresses on the human cranium might provide an explanation for the unusual morphology of WLH 50. His conclusion was that robusticity among ancient Australian remains is likely to have resulted from complex developmental interplays surrounding the unusual size of crania like WLH 50 – it is one of the largest human crania ever found – its shape – it is extremely long and narrow –, of crania like WLH 50, in combination with lifestyle and dietary factors in an arid zone dwelling hunting and gathering group and ageing processes (WLH 50 represents a male of advanced age)). That is, contemporary understanding of cranial growth and function provides a satisfactory and much simpler explanation for much of the variation we see among Pleistocene Australians without the need to resort to complicated models that invoke gene flow from ancient hominin populations.

This example demonstrates the fundamental role research into human variation continues to play in palaeoanthropology. Without knowledge of the likely expressions of variation (*normal*: individual, spatial and temporal: *abnormal* arising from pathology and/or developmental faltering), and their potential causes, it is impossible to properly assess models about population history and evolution; or to properly undertake taxonomic classification.

RELEVANCE OF TEACHING VARIATION

These discussions about some history themes of biological anthropology, importance of human variation to its various sub-disciplines and specific palaeoanthropological example provide materials that may be drawn upon in demonstrating the continuing importance of human variation. The present author also believes that the knowledge and skills base of anthropology are increasingly relevant to students in a world facing global challenges such as climate change and dramatic scientific and technological changes in fields such as medicine and information-communication technology. We also live in increasingly diverse, ever more interconnected, more ethically oriented workplaces and communities, and the study of anthropology provides opportunities to meet the needs of students living in such a world.

For example, projections of future changes in the Earth's climate have major implications for population movement, settlement and land use, as well as for human diet, disease and mortality (see e.g., McMichael, 2004). All of these issues have implications for understanding the future of human growth, development, variation and the genetic and physiological responses of our species to rapid climate and environmental change, on both short and long timescales. Graduates in anthropology with a solid background in human variation are well equipped to understand these changes and to contribute in a positive way professionally and socially to addressing such issues.

For example, with global warming, vector-borne diseases such as malaria and dengue fever are expected to enlarge their distribution into higher latitudes. This means that many populations will be exposed to such diseases for the first time, while others, perhaps exposed only seasonally, might well experience their consequences for much of the year. This is likely to result in both cultural and biological responses within such populations. Microevolutionary (genetic) changes are likely to occur as populations adapt, especially those without access to modern medical care for the treatment of malaria, which includes most people in the world today, and an increasing number of people in the future as climate change takes it economic toll on the developing world. Population frequencies for many genes will change in response to vector-borne diseases as people with molecular resistance (such as heterozygotes for sickle-cell anaemia) increase in numbers. Conversely, the severe anaemia suffered by homozygotes of conditions like sickle-cell will undoubtedly increase and add extra burdens to families, communities and health systems. Policy makers with a solid understanding of human variation will be well equipped to play a positive role in the future.

Research in anthropology is at the leading edge of medical, morphological, evolutionary and molecular science. For example, molecular anthropologists are leading research into identifying and understanding the function of genes associated with phenotypic and molecular variation between humans and other apes, and between people from different geographical regions. Some questions they address include: *How and why do humans differ from the other apes? Why are some human populations more susceptible to particular diseases than other groups?* Can medical therapeutics be designed to fit individual genetic profiles? The answers to these questions are leading to major developments in evolutionary anthropology, functional genomics, and developmental biology, and have even lead to the establishment of a new medical field: 'evolutionary medicine' (see e.g., Trevathan, 2007).

A Broader Curriculum

Over the last decade or so, Australian universities have followed a global trend of adopting and implementing generic capabilities, or graduate attributes, into their curricula. Bowden *et al.* (2000) defines such capabilities as:

> ...the qualities, skills and understandings a university community agrees its students should develop during their time with the institution. These attributes include, but go beyond, the disciplinary expertise or technical knowledge that has traditionally formed the core of most university courses. They are qualities that also prepare graduates as agents for social good in an unknown future.

This approach makes explicit for the first time the need to inculcate 'values' in education across all areas of university education and training; beyond the traditional vocationally oriented disciplines such as psychology, law, engineering and medicine. In particular, it requires university educators to actively engage with and promote ethical and socially responsible behaviours that are both societal/cultural (or generic) and discipline specific.

For example, the Faculty of Science at the University of New South Wales has developed six graduate attributes for its undergraduates to have acquired while undertaking its various bachelor's programs. One of these attributes illustrates how anthropology, particularly the study of human variation, provides a platform for successful integration of the graduate attributes into teaching:

> Ethical, Social and Professional Understanding. Ability to critically reflect upon broad ethical principles and codes of conduct in order to behave consistently with a personal respect and commitment to ethical practice and social responsibility. Understanding of responsibility to contribute to the community. Respect and value social, multicultural, cultural and personal diversity.

Anthropology offers many examples from both its history and its research subject matter to teach in a positive and empirically supported way the importance of ethical practice, social responsibility and social and physical diversity. It offers students the chance to reflect on real life, everyday encounters: we see and experience diversity every day whether within our families, social groups or broader community; and in terms of how we appear physically and the sorts of diseases we are likely to suffer. Understanding how and why people vary is crucial to dispelling stereotypes and myths surrounding ethnic and other social distinctions (such as class) and to removing the artifices of social division. An excellent example derives from anthropological research over the last 60 years or so demonstrating the powerful effect the environment has on children in both their physical and intellectual development; and on the related and well established secular trends in stature, body mass and reproductive maturity in many populations during the 20th century.

REFERENCES

American Association of Physical Anthropologists. (1996). AAPA statement on biological aspects of race. *American Journal of Physical Anthropology*. 101: 569-570.

Bowden, J., Hart, G., King, B., Trigwell, K., & Watts, O. (2000). *Generic Capabilities of ATN University Graduates*. Canberra: Australian Government Department of Education, Training and Youth Affairs. Available at http:/www.clt.uts.edu.au/ATN.grad.cap.project.index.html

Bowler, J. B., Johnston, H., Olley, J. M.. Prescott, J. R., Roberts, R. G., Shawcross, W. & Spooner, N. A. (2003). New ages for human occupation and climatic change at Lake Mungo, Australia. *Nature* 421: 837-840.

Curnoe, D. (2007). Modern human origins in Australasia: testing the predictions of competing models. *Homo - Journal of Comparative Human Biology* 58: 117- 157.

Curnoe, D. (2009). Possible causes and significance of cranial robusticity among Pleistocene and early Holocene Australians. *Journal of Archaeological Science* 36: 980-990.

Curnoe, D., & Thorne, A. (2006). Human origins in Australia: the skeletal evidence. *Before Farming* [on-line version] 2006/1 Article 5.

Frayer, D. W., Wolpoff, M. H., Thorne, A. G., Smith, F. H., & Pope, G. G. (1993). Theories of modern human origins: the paleontological test. *American Anthropologist* 95: 14-50.

Gould, S.J. (1988). *The Mismeasure of Man*. London: Penguin Books Ltd.

Hooton, E. A. (1926). Methods of racial analysis. *Science* 63: 1621.

Kohn, M. (1996). *The Race Gallery*. London: Vintage.

Lewontin, R.C. (1972). The apportionment of human diversity. *Evolutionary Biology* 6: 381- 398.

Lieberman, L., & Jackson, F.L.C. (1995). Race and three models of human origin. *American Anthropologist* 97: 231-242.

McMichael, M.J. (2004). Environmental and social influences on infectious diseases: past, present and future. *Philosophical Transactions of the Royal Society of London B-Biological Sciences* 359: 1049-1058.

Simpson, J. J., & Grün, R. (1998). Non-destructive gamma spectrometric U-series dating. *Quaternary Science Reviews* 17: 1099-1022.

Thorne, A. G. (1980). The longest link: human evolution in Southeast Asia and the settlement of Australia. In J.J. Fox, A.G. Garnaut, P.T. McCawley, & J.A. Maukic (Eds.), *Indonesia: Australian Perspective* (pp. 35-43). Canberra: Australian National University.

Thorne, A., Grün, R., Mortimer, G., Spooner, N.A., McCulloch, M., Taylor, L., & Curnoe, D. (1999). Australia's oldest human remains: age of the Lake Mungo 3 skeleton. *Journal of Human Evolution* 36: 591-612.

Trevathan, W.R. (2007). Evolutionary medicine. *Annual Review of Anthropology* 36: 139-154.

Relethford, J.H. (2002). Apportionment of global human genetic diversity based on craniometrics and skin color. *American Journal of Physical Anthropology* 118: 393-398.

Seligman, C.G. (1959). *Races of Africa* (3rd edition). London: Oxford University Press.

Webb, S.G. (2006). *The First Boat People*. Cambridge: Cambridge University Press.

In: Teaching Human Variation
Editor: Goran Štrkalj
ISBN: 978-1-60876-616-1

Chapter 2

RACIAL IDENTIFICATION OF SINGLE SKULLS IN FORENSIC CASES: WHEN MYTH BECOMES REALITY

Alan G. Morris[*]
Department of Human Biology, University of Cape Town, Cape Town South Africa

ABSTRACT

Forensic identification software such as Fordisc 2.0 use multivariate statistics to place a cranium within a parental 'racial' population according to its morphology. Although widely used, this technique is only really appropriate when the unknown cranium is drawn from one of the 'parent' populations used in the original calculation of statistical algorithms. The software programme is therefore unreliable when used on unknown persons drawn from different 'parent' samples (Williams *et al.* 2005).

The idea that humans can be conveniently categorised into neat racial groups has been with us for a very long time, and is widely believed by a large number of scientists. This is nowhere more obvious than in South Africa, where the racial classification of its citizens into 'White', 'Black', 'Coloured', and 'Indian' is still commonly accepted despite the dismissal

[*] E-mail: Alan.Morris@uct.ac.za

from government of the original creators of the apartheid system. That these categories are essentially political and/or cultural and that there is significant morphological overlap between categories is almost completely missed, and the racial categories continue to be used by police investigators and government agencies as distinct and non-overlapping groupings.

This paper examines a set of measurements from a range of South African populations and shows how the actual nature of the data can be used to understand patterns of cranial morphology in the region. The difficulty in using these data for forensic identification will be emphasised.

INTRODUCTION

The May 2009 volume of the *American Journal of Physical Anthropology* published a set of papers originally presented at a symposium entitled "Race reconciled?: How biological anthropologists view human variation", held in May 2007 in New Mexico. The delegates agreed that the old typological race concept was not a very accurate way of describing human variation (Edgar & Hunley, 2009), but there remained a considerable difference of opinion between the morphologists and the geneticists as to whether any aspect of the race concept could be salvaged. Lineage tracking by DNA markers clearly show modern humans to be a subset derived from African variants (Long *et al.*, 2009) and that the observed pattern of genetic variation does not support a simple three group racial categorisation. This tracking, in particular, indicates that genetic lineage identity does not vary in direct relationship to geographic distance and that lineage is the result of a complex history of bottlenecks and rapid long distance migrations from an African core (Hunley *et al.*, 2009). In opposition, Relethford (2009) eloquently demonstrated that despite the lineage data from genetics, world-wide geographic patterning in skin colour and cranial form produce observable regional differentiation but that this is clinal in distribution and all racial categories do is transform this continuous variation into discontinuous and fairly distinct clusters. Relethford was careful not to imply that this is the basis of a rigid classification, but he noted that cranial shape is a good reflection of regional variability and that statistical classification techniques does have high success rates.

Relethford is not a forensic specialist, but he acknowledges that cranial shape has the potential also to be useful in forensic contexts "where one might simply want to assign a specimen to a broad ancestral group" (Relethford, 2009: 21). This is indeed the technique of choice for forensic anthropologists. The most important tool in identifying these 'ancestral groups' is FORDISC, the discriminant function analysis software distributed by the University of Tennessee (Ousley & Jantz, 1996). FORDISC's associated data base contains the original Howells' world-wide data set (Howells, 1973, 1989), plus an extended 'Forensic Data Base' of American forensic cases (Ousley *et al.*, 2009). The use of this technique is now widespread in North America, but there is some doubt about its appropriateness for samples outside of the reference collections (Williams *et al.*, 2005).

This paper examines the specific case of South Africa in relation both to the reality of racial grouping in the country and to the use of FORDISC and other multivariate assessment methods in assessing population affinity of single skulls in the local context. The central issue is that what seems logical and consistent in science is not necessarily understood the same way by the lay public. Two aspects of education are involved: informing the public about the limitations of the classification system, and educating the scientists about how to apply the classification in the context of both social and biological variation.

Statistical Identification of Race using Cranial Morphology

The important work of Howells on cranial variation (1973) makes use of Principal Components Analysis to differentiate data from 28 samples of human crania selected by Howells between 1963 and 1965. Once these 'reference samples' were quantified, it was a relatively easy task to use Discriminant Analysis to sort unknown individuals into one or another group. Simply put, the Discriminant technique as used in programmes such as FORDISC essentially creates a bias in an analysis to load measurements that enhance differentiation so that maximum divergence between reference groups is obtained. An unknown individual is then drawn toward the group mean which it most closely approximates and the probability of its membership in that group is then calculated.

Williams *et al.* (2005) attempts to test the Principal Components Analysis technique using FORDISC to analyse a set of prehistoric Nubian crania. The results are surprising in that relatively few of the archaeological crania cluster with the Howells 'Egyptian' sample to which they are most closely related archaeologically and geographically. Many of the 'incorrectly' assigned individuals associate with reference samples originating far away from north-east Africa. Ousley *et al.* (2009) respond to this critical paper with a full re-analysis of the original data set and they argue that the problem is in the way Williams *et al.* did their research rather than in the technique itself. Ousley et al. demonstrate that when the Nubian crania are differentiated based on the same 10 cranial measurements used by Williams *et al.*, 50% of the skulls are correctly classified as Egyptian and a further 20% are classified as coming from other reference samples in north-east Africa. When the number of measurements is increased to 24, almost 75% of the Nubians are correctly classified with Egyptians and nearly 90% are identified as coming from a group in the same region. Ousley *et al.* (2009) then go on to note that the identical analysis of craniometric variation is extremely accurate in differentiating between American blacks and whites.

At the same conference in 2007, Konigsberg *et al.* (2009) presented a report of a single individual from a forensic case in Iowa using 11 cranial measurements with a FORDISC analysis. The unknown individual was classified as coming from Easter Island (one of Howells' reference samples) despite the fact that it was patently obvious that it was extremely unlikely to be correct. They concluded that the problem was in the question, not the analysis. The technique assumes racial homogeneity and works well when separating 'blacks, whites and Asians', but does not work well when a full range of variation is considered.

Nevertheless, interest in application of the technique continues and two papers examining FORDISC analysis were presented at the American Association of Physical Anthropology meeting in Chicago in March 2009. Aronsen & Ellis (2009) analysed 13 known crania from a collection at the Yale Peabody Museum of Natural History. The skulls originated from a wide range of geographic regions. The FORDISC programme was successful in associating 9 of the skulls with their original parent population using the Forensic and Howells' data bases, and the authors recommended that the reference sample choices be reviewed as the system does have errors. Subsequently, Elliot & Collard (2009) reanalyzed the original Howells data set using 56 variables and then analyzed it again with only 10 variables. They concluded that the sex of each specimen needs to be known in advance and that any analysis should have more than 10 variables, but that "even when these conditions are met, few determinations are likely to be unambiguously correct" (Elliot & Collard, 2009:124)

These critiques clearly show that there are problems in applying the FORDISC approach to populations outside of the FORDISC database. In my view, much of the problem in using the FORDISC analysis originates from the discriminant technique itself. The results of the analysis always links the unknown to its nearest group mean, no matter how tenuous the association, as long as it is stronger than the other links (Howells, 1970). Because of this, the unknown should be a member of one of the reference samples (Corruccini, 1975) for more accurate results. Increasing the number of variables helps to boost accuracy, but even then the technique works best only when the search categories are limited to geographically and morphologically distinct categories.

The first attempt to apply the FORDISC approach in the South African context is by L'Abbe *et al.* (2008). These authors applied the most recent version of FORDISC to 226 individuals drawn from the cadaver collection of human skeletons at the University of Pretoria, 15 unspecified 'Griqua' individuals, and 24 late 19th century African farm labourers whose skeletons were from the National Museum collection in Bloemfontein. The Griqua, an historic 'coloured' community from the northern Cape Province, were presumably added to include a 'mixed race' group. Fifteen measurements were used in the analysis. Overall, 73% of crania were correctly assigned to sex and ancestry, with accuracy the best for white females (82%) and the worst for black females (60%). The errors in assignment were interpreted not as mistakes resulting from the statistics of the programme itself, but rather, occurring because the FORDISC datasets are not representative of the South African population.

Racial Variation and Forensic Identification in South Africa

The work by L'Abbe *et al.* (2008) applies the FORDISC database directly to South African individuals, but essentially poses its question of identification with only two main categories—black and white. The underlying assumption is that the crania are drawn from the 'white' population (donations to the medical school) or from 'black' population (either from paupers or from the farm workers whose skeletons are housed at the museum in Bloemfontein). The inclusion of the Griqua skeletons in the sample presumably adds a sample of people who have a mixed ancestry, but this is a poor assumption as the genetic origins of the Griqua are extremely complicated (Morris, 1997) and this aspect is not discussed in the poster presentation. In fact, much of the work on forensic identity coming from the Pretoria school has rigid dichotomy of black and white (for example see Patriquin *et al.,* 2002; Patriquin *et al.*, 2003). Part of this restricted view is caused by a collaboration with American researchers who import the American perspective of rigid racial views, but it also reflects a social reality of South Africa's divided society as it appears in the large cities of Gauteng Province. Research coming from the school in Cape Town is quite different, and racial identification there is seen as a much less important research focus. Research in the Cape either does not include biological ancestry (see for example Morris' 1992 (a) catalogue where race categories are entirely absent and only geography and time or archaeological association are considered), or race is seen as a self-identified category not generated by the researcher (Freidling & Morris, 2005). Even standard forensic reports differ at the two centres. In Pretoria, categorization of race is a distinct part of standard biological assessment, while in Cape Town emphasis is placed on using features to exclude membership in specific categories, rather than using those same features to define membership in one group or other. In essence, this procedure has the same effect, 'reducing search categories', but it is philosophically quite different.

Why are the two schools different? The answer to this question is perhaps more obvious in Cape Town than in Pretoria, but both schools are influenced by the underlying regional 'Folk' rather than 'Scientific' categories of people. A 'Folk Taxonomy' is the popular perception of racial categories within a society. Although it may have some of its roots in actual biological differences between people, a Folk taxonomy is usually characterised by the conflation of ethnicity and biology, and it is generally unstable, changing with the social and political transformation of the society (Edgar, 2009). However, it is the folk taxonomy that can have a real impact on people's lives, especially as related to social and health inequalities. In Apartheid South Africa (1950 to 1991), the folk taxonomy was reified by law and impacted every aspect of every citizen's life. Although legislated Apartheid is long gone, it's system of classification still has great influence (Seekings, 2008).

The Apartheid classification of the mid-20th century was intended to entrench the economic and political dominance of the white minority, and also to ensure that the white group maintained an imagined racial purity and was not 'diluted' by intermixture with other races (Seekings, 2008). An original system of three races (White, Bantu & Mixed or 'Coloured') evolved into a four category administrative system (White, Black, Coloured and Indian), but eventually these categories were subdivided into a myriad of smaller groups defined primarily by culture. These included separate Black identities for each 'Homeland' which had residential and political implications, and separate identities for various coloured and Asian groups (e.g., Malay, Indian, Chinese, Griqua), which had social and residential ramifications. Although there were obvious ethnic differences within the white group, these were never administratively separated.

The end of Apartheid officially brought an end to this system of classification and from 1991 race is no longer a category entered into the Population Register. However, most South Africans continue to have clear racial identities and view others in racial terms. There is still very little social or residential integration. South Africans are slowly beginning to identify themselves in terms of a national "South African" entity (Friedling & Morris 2005; Seekings, 2008), and this is resulting in class and other social-cultural categories becoming more important than race, but the continuing economic inequity prevents a clean break of the link between race and class.

Seekings (2008) points out an obvious divide in racial perception between Whites and others in South Africa. Black and Coloured South Africans tend to think of themselves in terms of 'culture' groups, whereas Whites tend to perceive physical features such as 'paler skin' to be more important. Culture for Whites appears to be less of an issue than biological appearance. Coloured people have tended to internalise a negativity about their identity as being 'neither white nor black'; nevertheless, the idea of a Coloured identity separate from other groups has strengthened since the end of Apartheid because of a perceived loss of privilege of Coloured people to Black African members of society (Adhikari, 2006). It is very much this resurgence of Coloured identity that influences the view of race in Cape Town as opposed to Pretoria.

The historical gene flow between settlers and Africans in Cape Town has resulted in a group of people who are identified as, and sometimes self-identify as "Cape Coloured". This identification is the result of a long historical process, and although the name implies some form of homogeneity, the reality is spectacularly complex and cultural or biological uniformity is impossible to demonstrate. In one of the rare studies that concerns itself with the genetic origins of these people, Tishkoff *et al.* (2009) indicate that from a DNA perspective, the Cape Coloured group has the "highest levels of intercontinental admixture of any global population," which is something historians have known for a long time (Shell, 1994).

For students of cranial morphology, this observation means that Cape Town displays a continuous representation of world-wide cranial shape and that analysis of structure will not allow easy categorisation of unknown individuals. But even more important is the fact that the folk category 'Coloured' does not reflect any definable biological category. Despite the wishes of the Apartheid administrators, gene flow was continuous between the old 'White', 'Coloured' and 'Black' groups, and a simplistic interpretation of the Cape population into three race groups according to physical appearance is not viable. In addition, there are genetic eddies within the 'Coloured' monolith that are poorly understood and need research. Assortive mating based on cultural markers such as religion is strong, and history has determined that the formation of 'Coloured' people in urban Cape Town is different from that in the nearby and more distant rural areas.

Historic data for 18th century Cape Town suggest that at least half the population at that time was enslaved or otherwise excluded from white settler society (Shell, 1994). The vast majority of the enslaved people originated elsewhere in Africa or in Dutch colonial lands in south-east Asia, but significantly missing were large numbers of Khoesan people. The situation in the country-side was quite different. Barrow, traveling in the Graaff-Reinet district in the last years of the 18th century, noted that there were only about 600 to 700 slaves in the whole district, but there were at least 10,000 'Hottentots' (Eldredge, 1994). This suggests that the Khoesan people were the largest demographic group in the countryside, but almost missing from the urban situation. The demographic situation in Cape Town changed with the end of slavery in the 1830s, when many Khoesan people from the rural districts moved to the city to take advantage of new economic opportunities opened up by the end of slave labour. Modern 'Coloured' populations are therefore strongly shaped by historic events of the 18th and 19th century, and continued to be shaped by political events in the 20th century.

A Case Study: The Cobern Street Burials

The historic burial ground at Cobern Street in Cape Town was discovered by property developers in September 1994. No formal records of the burial ground are available, but archival records suggest that this site is a cemetery allocated to slaves and "free black" peoples of Cape Town in the late 18th century. In total, 65 burials and a number of bone scatters were recovered at the site.

Cranial and mandibular measurements were recorded for the 25 more or less complete skulls from the Cobern Street series. The measurements are grouped by calvaria, face and mandible in order to maximise the number of less than complete individuals for analysis. The multivariate method of analysis used is Principal Components instead of Discriminant Analysis, as the object of the study was not classification but rather an analysis of genetic dispersion. The reference samples were selected to try to reflect some of the known genetic inflow during the historic period at the Cape, which includes Europeans (White South Africans), Khoesan, Bantu-speaking South Africans, South Indians and in the case of the males, Chinese. No sample of Indonesian origin was available for study.

Figures 1 and 2 display the dispersion of individuals for the analyses of male and female, respectively. The object in this paper is not to discuss the specific Cobern distribution, but to use the series as an example to show the difficulty of identifying single individuals.

In the same manner observed in Discriminant analyses (Morris, 1992b), the reference samples of South African Whites, Blacks and Khoesan differentiate reasonably well, demonstrating their distinct morphological differences. In general, the Cobern sample takes a middle position in the dispersion, but overlaps with all groups to some degree.

Much more important is that not only is the Cobern sample heterogeneous, but each of the reference samples also has a wide dispersion on the graph. Although it is possible to see patterns in the dispersion, it is obvious that single individuals cannot easily be assigned to any specific reference group based on their morphology and added to this is the fact that the reference samples are only an approximation of the original parental populations.

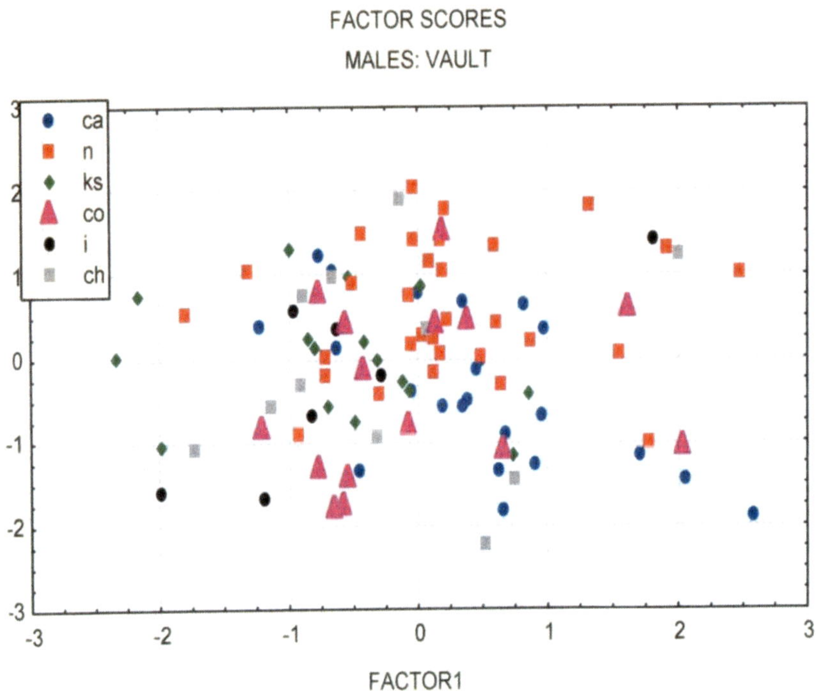

9 Variables, 56.7% of variation in Factors 1 and 2.

KEY:

ca = South African whites
n = South African blacks
ks = Khoesan
co = Cobern Street
i = South Indians
ch = South Chinese

Figure 1a. Principal Component Distribution for Male Vaults.

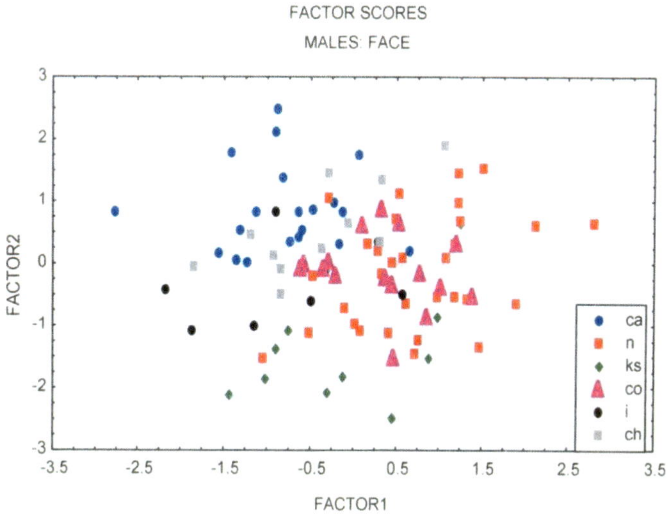

16 Variables, 51.8% of variation in Factors 1 and 2. See figure 1a for sample labels.

Figure 1b. Principal Component Distribution for Male Faces.

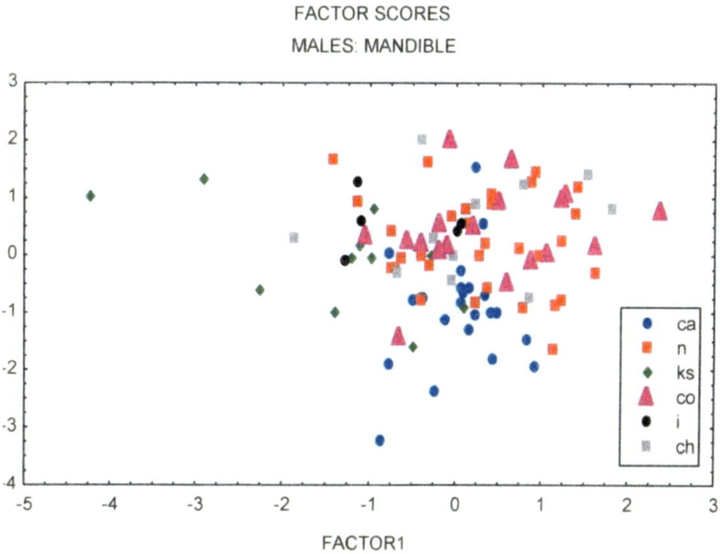

9 Variables, 53.2% of variation in Factors 1 and 2. See figure 1a for sample labels.

Figure 1c. Principal Component Distribution for Male Mandibles.

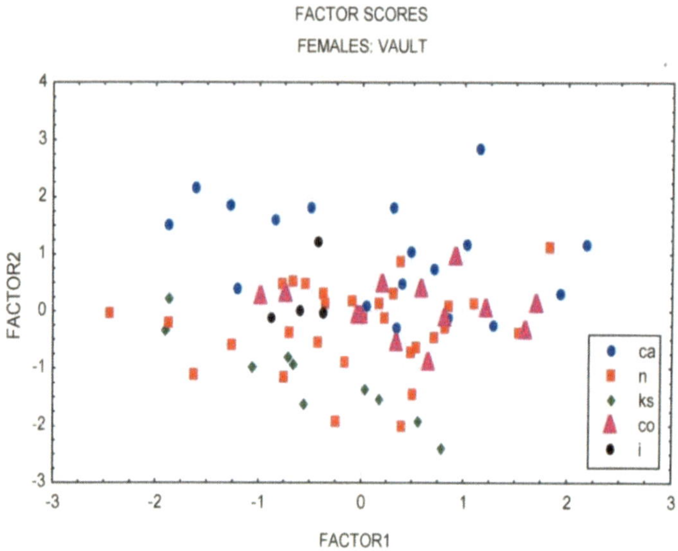

12 Variables, 53.1% of variation in Factors 1 and 2.
See figure 1 a for sample labels

Figure 2a. Principal Component Distribution for Female Vaults.

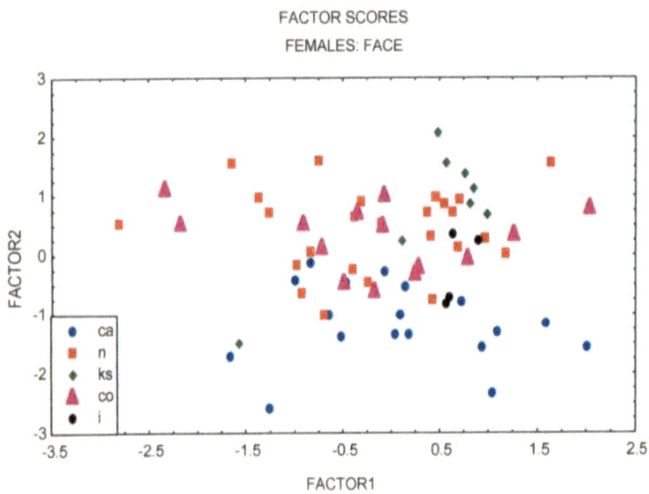

10 Variables, 59.6% of variation in Factors 1 and 2.
See figure 2 a for sample labels.

Figure 2b. Principal Component Distribution for Female Faces.

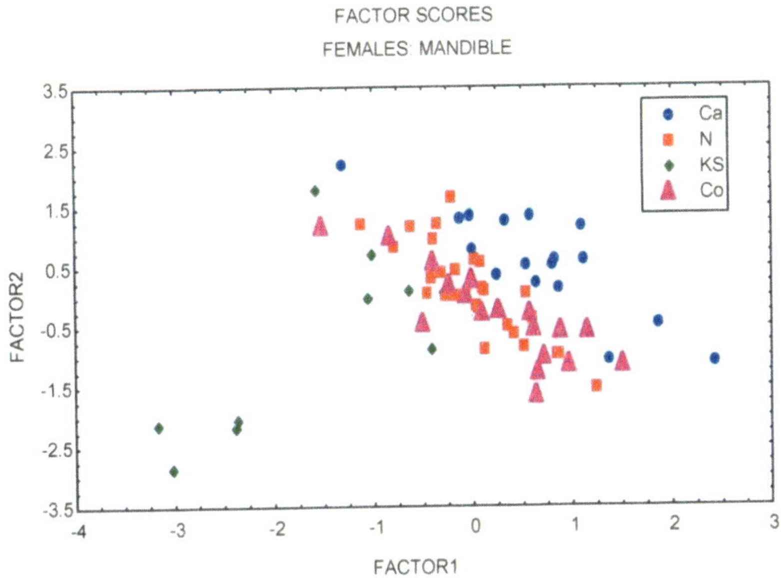

9 Variables, 53.9% of variation in Factors 1 and 2. See figure 2 a for sample labels.

Figure 2c. Principal Component Distribution for Female Mandibles.

The importance of the Cobern Street site remains is that they provide us with a window into the biology of the common people of Cape Town during the latter part of the Dutch occupation. This biological picture becomes even more complex in the succeeding two centuries, yet the Folk taxonomy of the "Coloured" identity remains a single entity in the view of most citizens today.

Lessons for Forensic Identification of Single Skulls

The only reason forensic studies continue to use racial categories is to help to determine the specific identity of individuals. Ousley *et al.* (2009) amongst others argue that the discriminant techniques work and that it is possible to identify individuals of different racial origins with confidence, but a growing number of researchers express doubt in that confidence. The FORDISC reference samples are selected to be as different as possible from each other so that group identification is very clear, but considerations of morphological overlap between groups is conveniently forgotten. L'Abbe *et al.* (2008) do not fault the analytical technique, but imply that the FORDISC programme would work better if the reference samples were specifically South African. To some degree they are right, but even with enhanced samples, linking the Folk taxonomy of usage in police circles to scientific classification of the skulls would remain problematic. Not only is it unwise to assume a simple Black versus White dichotomy in the populations of Pretoria and Johannesburg, but the influx of individuals from other parts of Africa is now so pronounced that even a reference sample of South African Blacks might be inappropriate.

But am I perhaps being overly critical? Is it not reasonable to study cranial variation of South Africans in order to understand their anatomical variability? Actually, such studies are not unreasonable and need to be done, but it might be more useful to understand the extent of the variation as the central object of study. With such extensive morphological overlapping between populations as evident in the, it becomes more important to see if given individuals are outside of the normal range of variation rather than 'typical' for the reference sample. The exclusion principle is easier to apply with confidence than trying to argue a specific identity when different identities with similar morphologies are possible.

Forensic anthropologists also must think very clearly about why they ask the race question at all. Novotńy *et al.* (1993:76) argue that "the identification of unknown individuals is one of the most important justifications for maintaining biologically based racial typologies". But in fact, the identification of an individual is NEVER made on the basis of his or her race. The final identification is done based on the distinct appearance of the individual. Visible features which mark us as individuals are NOT related to race. At best, the race identification is capable of reducing the number of search categories. In both Cape Town and Pretoria, police continue to use racial categories based on the old Apartheid system for identification; thus, confusion might still arise if a forensic anthropologist makes a formal determination of race by examining a skull, when in fact, the actual identity of the living person is determined by the mixed bag of physical, linguistic and cultural characters that all of us use to determine our own identity.

Edgar & Hunley (2009) lament that American students still hold strong typological views of race because anthropologists fail to offer a clear and satisfactory explanation that meshes with students' life experience. The same is true for the South African context although the details differ. In South Africa, the problem of racial identity in forensic cases is first and foremost about how South Africans perceive racial identity. Applications of discriminant programmes such as FORDISC are mathematically correct, but the categories and reference samples are researcher selected, and may not directly overlap with the Folk taxonomy (i.e., perceived racial identity)_ used by the police and understood by the court officials in South Africa.

ACKNOWLEDGEMENTS

I thank Isabelle Ribot, now of the University of Montreal, for her expertise in analysing the Cobern Street skeletons. I would also like to give recognition to the excellent research in forensic anthropology by my Pretoria colleagues despite the fact that I may disagree with specific aspects. Perhaps there is room here for a mingling of minds between the two centres.

REFERENCES

Adhikari, M. (2005). *Not White Enough, not Black Enough: Racial Identity in the South African Coloured Community*. Ohio University Research in International Studies, Africa Series Number 83. Athens: Ohio University Press.

Aronsen, G.P., & Ellis, S.C. (2009). Identification of unprovenenced human remains using FORDISC 3.0: biological race revisited. *American Journal of Physical Anthropology* Supplement 48: 79-80.

Corruccini, R.S. (1975). Multivariate analysis in biological anthropology: some considerations. *Journal of Human Evolution* 4: 1-19.

Edgar, H.J.H. (2009). Biohistorical approaches to "race" in the United States: biological distances among African Americans. *American Journal of Physical Anthropology* 139: 58-67.

Edgar, H.J.H., & Hunley, K.L. (2009). Race reconciled?: How biological anthropologists view human variation. *American Journal of Physical Anthropology* 139: 1-4.

Eldredge, E.A. (1994). Slave raiding across the Cape frontier. In E.A. Eldredge, & F. Morton, F. (Eds.), *Slavery in South Africa: Captive Labour on the Dutch Frontier* (pp.93-126). Pietermaritzburg: University of Natal Press.

Elliot, M., & Collard, M. (2009). FORDISC and the determination of ancestry from craniometric data. *American Journal of Physical Anthropology* Supplement 48: 124.

Friedling, L.J., & Morris, A.G. (2005). The frequency of culturally derived dental modification practices on the Cape Flats in the Western Cape. *Journal of the South African Dental Association* 60(3): 97-102.

Howells, W.W. (1970). Multivariate analysis for the identification of race from crania. In T.D. Stewart (Ed.), *Personal Identification in Mass Disasters* (pp. 111-121). Washington: Smithsonian Institution.

Howells, W.W. (1973). *Cranial Variation in Man*. Papers of the Peabody Museum of Archaeology and Ethnology, Vol. 67. Cambridge:Harvard University.

Howells, W.W. (1989). *Skull Shape and the Map: Craniometric Analyses in the Dispersion of Modern* Homo. Cambridge, Mass.: Harvard University Press.

Hunley, K.L., Healy, M.E., & Long, J.C. (2009). The global pattern of gene identity variation reveals a history of long-range migrations, bottlenecks, and local mate exchange: implications for biological race. *American Journal of Physical Anthropology* 139: 35-46.

Konigsberg, L.W., Algee-Hewitt, B.F.B., & Steadman, D.W. (2009). Estimation and evidence in forensic anthropology: sex and race. *American Journal of Physical Anthropology* 139: 77-90.

L'Abbe, E.N., Nawrocki, S.J., & Keough, N. (2008). The application of Fordisc 3.0 to South African crania. Poster presented at the University of Pretoria Health Sciences Faculty Day, August 2008.

Long, J.C., Li, J., & Healy, E. (2009). Human DNA sequences: more variation and less race. *American Journal of Physical Anthropology* 139: 23-34.

Morris, A.G. (1992a). *A Master Catalogue: Holocene Human Skeletons from South Africa.* Johannesburg: Witwatersrand University Press.

Morris, A.G. (1992b). *The Skeletons of Contact: Protohistoric Burials from the Lower Orange River Valley.* Johannesburg: Witwatersrand University Press.

Morris, A.G. (1997). The Griqua and the Khoikhoi: biology, ethnicity and the construction of identity. *Kronos* 24: 106-118.

Novotńy, V., İscan, M.Y., & Loth, S.R. (1993). Morphologic and osteometric assessment of age, sex and race from the skull. In M.Y. Iscan, & R.P. Helmer (Eds.), *Forensic Analysis of the Skull* (pp. 71–81). New York: Wiley-Liss.

Ousley, S., Jantz, R., & Freid, D. (2009). Understanding race and human variation: why forensic anthropologists are good at identifying race. *American Journal of Physical Anthropology* 139: 68-76.

Ousley, S., & Jantz, R. (1996). *Fordisc 2.0: Personal Computer Forensic Discrimination Functions*. Knoxville: University of Tennessee.

Patriquin, M.L., Steyn, M., & Loth, S.R. (2002). Metric assessment of race from the pelvis in South Africans. *Forensic Science International* 127: 104-113.

Patriquin, M.L., Loth, S.R., & Steyn, M. (2003). Sexually dimorphic pelvic morphology in South African whites and blacks. *Homo* 53: 255-262.

Relethford, J.H. (2009). Race and global patterns of phenotypic variation. *American Journal of Physical Anthropology* 139:16-22.

Seekings, J. (2008). The continuing salience of race: discrimination and diversity in South Africa. *Journal of Contemporary African Studies* 26(1):1-25.

Shell, R. (1994). *Children of Bondage*. Johannesburg: Witwatersrand University Press.

Tishkoff, S.A., Reed, F.A., Friedlaender, F.R., Ehret, C., Ranciaro, A., Froment, A., Hirbo, J.B., Awomoyi, A.A., Bodo, J-M., Doumbo, O., Ibrahim, M., Juma, A.T., Kotze, M.J., Lema, G., Moore, J.H., Mortensen, H., Nyambo, T.B., Omar, S.A., Powell, K., Pretorius, G.S., Smith, M.W., Thera, M.A., Wambebe, C., Weber, J.L., & Williams, S.M. (2009). The genetic structure and history of Africans and African Americans. *Sciencexpress* www.sciencexpress.org/30April2009/Page1/10.1126/science.1172257

Williams, F. l'E., Blecher, R.L., & Armelagos, G.J. (2005). Forensic misclassification of ancient Nubian crania: implications for assumptions about ancient human variation. *Current Anthropology* 46:340-345.

In: Teaching Human Variation
Editor: Goran Štrkalj
ISBN: 978-1-60876-616-1

Chapter 3

THE ILLUSIVE CONCEPT OF HUMAN VARIATION: THIRTY YEARS OF TEACHING BIOLOGICAL ANTHROPOLOGY ON FOUR CONTINENTS

Maciej Henneberg[*]
Biological Anthropology and Comparative Anatomy Research Unit,
University of Adelaide, Adelaide, Australia

ABSTRACT

Biological anthropology largely fails to provide a coherent and generally accepted approach to practical describing of human morphologies in terms of simple categories. This is not to say that biological anthropologists do not have good understanding of the complexity of human variation within and across populations. Their explanations, for whatever reasons, are not accepted by popular culture or other disciplines.

In biological anthropology, it is well recognised that the majority of variation in human biological characteristics occurs amongst individuals of the same sex in the same population, while sexual dimorphism and interpopulation variation constitute each about one-quarter of the variation of most morphological characters. Thus, it is practically

[*] E-mail: maciej.henneberg@adelaide.edu.au

impossible to reliably assign an individual to a particular population based on a cursory observation of a few external characteristics. Multivariate discriminant functions perform somewhat better, but accuracy of assignment is far from certain. Moreover, continuously acting microevolutionary forces can quickly change variation patterns. Thus, biological anthropologists possess a sound argument against categorisation of human beings into races, types or subspecies. However, denying the existence of patterns that people, however incorrectly, are taught to recognise from childhood with regard to categorisation of other humans is not successful. Thus, biological anthropologists fail to oppose racial classifications.

This failure of biological anthropology leads to the practical use of categorisations of humans based on the cultural traditions of their particular societies, often without sound scientific basis: thus, folkloric concepts of human races or groupings arise. These practices of categorisation can be both biased and harmful. A biological anthropologist teaching university students who grow up in a particular society, and who are exposed to the daily use of racial categories in the media and among friends, faces the difficult task of explaining the actual state of human variation. Opposing an overwhelming cultural practice by criticising it, without offering a practical alternative, does not work. Students give politically correct answers in the biological anthropology course examinations and then continue to use locally, culturally accepted terms and concepts in their practical lives.

Worse still, this practice of using locally, culturally accepted terms and concepts is accepted as reality by forensic scientists who, responding to pressures from police, determine the "race" of victims and suspects they are helping to identify.

This author has taught biological anthropology and human biology on four continents, in differing cultural systems, and describes here differences in the practical approach to categorising human variation amongst his students, despite all of them being given the same explanation of human variation in his lectures.

INTRODUCTION

Each human being is different. Yet, each human being perceives reality in terms of categories. Hence the problem: We cannot have as many categories of human beings as there are individuals alive; however, any attempt to distribute human continuous variation into distinct categories is inadequate because it requires simplifications that some individuals or groups find objectionable. This dilemma arises from the fact that human intergroup variation is small compared with variation within a group.

The majority of variation of human biological characteristics occurs amongst individuals of the same sex in the same population. Sexual dimorphism and interpopulational variation constitute each only about one-quarter of the variation of most morphological traits. Thus, it is practically impossible to reliably assign an individual to a particular population based on a cursory observation of a few external characteristics. Multivariate discriminant functions perform somewhat better, but accurate assignment is far from certain.

In the last few centuries, the process of colonisation, and most recently of globalisation, brought together people from communities that for many previous centuries were relatively isolated from each other. Thus, an interposition of fair skinned Europeans against dark-skinned Aboriginal Australians produced the impression of a dichotomy of human types in Australia. Similarly, the importation of African slaves to the Americas produced the illusion of a simple categorical distinction between "races". Different lifestyles, chosen freely, or enforced by slavery, strengthened the perception of these dichotomies and embedded them into everyday thinking. Indeed, the tendency to categorise humans into simple, preferably dichotomous, types seems to be universal. In countries of Central and Eastern Europe, that remain free from large-scale importation of people from other continents, minor physical and cultural traits are used to categorise people: the concept of the "Nordic race" as opposed to the "Mediterranean", and the singling out of Jewish Europeans as separate biological entities are but two examples. We also have an unfortunate tendency to attach different values or social status to categories once created. In this way races, however poorly defined, have become a tool of reasoning and of decision making in social encounters: "us" and "them", "higher" and "lower" seem to indicate who we should like and whom to avoid.

It is difficult for teachers, especially when dealing with young adult people, to effectively change patterns of thought engrained in their minds from the early childhood. This situation poses a problem for biological anthropologists trying to explain the distribution of human variation and the futility of intra-species classifications of humans. In my view, we seem to be nay-sayers: That is, we opine that there are no races, no differences between groups of people, instead of replacing a poorer mental construct with a better one. It is difficult to change the mental pattern of categorical distinctions into a pattern of no distinctions, which is alien to the way humans think.

The Pattern of Learning about Human Variation

Students remember what they learn better when they are given an explanation fitting into their already existing thought patterns. This is why teachers often use metaphors when introducing new subjects. The famous metaphorical use of butterflies to teach about sexual reproduction has parallels in other subjects.

Growing children from early on learn to recognise and categorise human beings they encounter. These classifications are initially based on simple "nice" and "bad", but as the number of people a child comes into contact with increases, more sophisticated systems of classification are needed. The common one is "us" and "them" based on language, ethnic (cultural) designation, and physical appearance. The physical appearance often also includes hairstyle, facial hair, and clothing and adornments. Parents are important in teaching these classifications. Primary schooling systems may reinforce these categorisations through teaching of languages and simple cultural geography. Americans are our friends, while X are dangerous. In schools, there are usually just two categories of language: "ours" and "foreign", i.e., "theirs". The latter may be interesting, but it is not a real one. Similarly, the clothing children wear is either "appropriate" or not, unless there are school uniforms. In the popular culture of young people, certain casual dress styles and kinds of body adornment (tattoos, jewellery) are either "cool" or not.

Early categorisations of people are taught to allow children some basis for decision-making in order to ensure their safety. These early teachings, however, become deeply embedded, and they are difficult to change via additional teaching, up to the university level. They are also reinforced in informal social contacts and through the media.

In various societies, there are different common, public, systems of categorisation of people. They are needed to be able to cope with the daily reporting of events and description of situations. All these categorisations are oversimplified and are often based on historical, subjective, perceptions of human variation. They vary from country to country depending on each one's history of migration and relations with neighbours. In biologically and culturally homogenous societies, any foreign-looking person is an object of curiosity, while in mixed societies, differences become a marker used in social and economic competition. Such markers are based on specific local histories and may include a particular form of dress, physical appearance, accent or religious denomination.

TEACHING BIOLOGICAL VARIATION IN HUMANS

One of the major tasks of biological anthropology (Czekanowski, 1965) is to rationally explain to people why their compatriots, neighbours and other people look different from themselves, how these differences came about and what meaning they have, if any. As biological anthropologists, we all agree that all humans alive at present on the Earth are members of the same biological species, that this species is not subdivided strongly enough to distinguish within it various sub-species or different breeds and that local populations differ slightly in the frequency of various genes, some of which produce noticeable differences in physical appearance, body shape and size (Brace, 2005). These interpopulational differences are transient and alter relatively quickly through microevolutionary processes (Henneberg, 1997, 2006; Solomon *et al.,* 2009). Beyond this common core of knowledge, the opinions of various anthropologists as to the notion of race differ (Štrkalj & Gibbon, 2007). A breakdown of human variation for body size and brain size is illustrated in Fig. 1 indicating relatively minor share of variation among populations in the total variation.

Since 1973, I have been teaching biological anthropology at universities in Poland, the USA, South Africa and Australia. Although the understanding of human biological variation by biological anthropology is the same worldwide, different cultural/historical backgrounds and different realities of human variation in each country produce separate challenges to my university teaching.

In Poland, students showed very little disbelief in what we taught and quite a good understanding of microevolutionary processes shaping human variation. The population of Poland in 1970-1980 was biologically relatively homogenous; occasional migrants from other continents were rare and were usually welcomed. In the society at large "race" did not play any important role, in economic opportunities, in political life or in social/cultural interactions. Significant differences in those spheres were rather based on religious/political beliefs or on traditional social status. Racial terminology exists in the Polish language, but terms denoting a person's physical appearance have little, if any, emotional loading in daily discourse, nor are they used in political disputes.

In Texas in the mid-1980s, a discrepancy existed between what we taught and what students wrote back to us in examinations, and everyday life. That society remained strongly segregated along racial lines, in terms of residential areas and socio-economic status. However incorrect, the differences between communities defined largely by the skin colour, reached far into the socio-cultural sphere of the general population, becoming "ethnic differences". While it is true that the impact of legal equality and generally liberal intellectual attitudes on campus were strong enough to produce no visible overt discrimination at the University of Texas at Austin, across the freeway from its campus, in the black residential area, a separate university (Huston-Tillotson University) is attended virtually exclusively by African-American Texans. The realities of social distinctions between Black and White were preserved, separate from the information put forth in official learning, in class. It is difficult to tell what impact my teaching had on the attitudes and opinions of my students after they graduated and entered the real world.

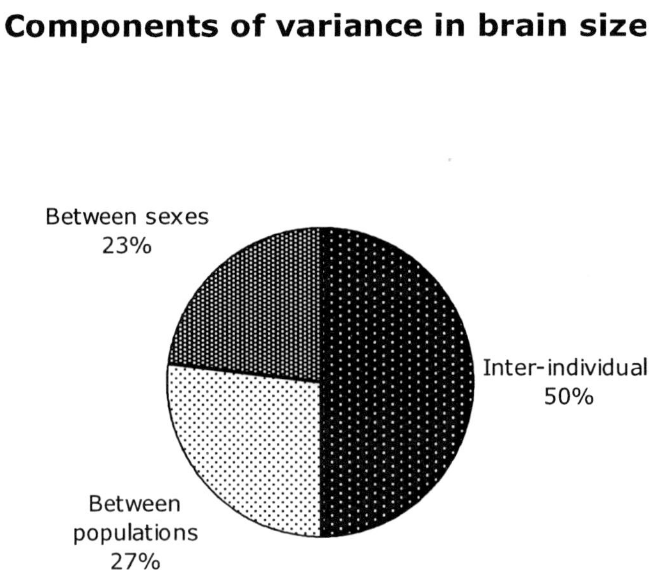

Figure 1. Continued on next page.

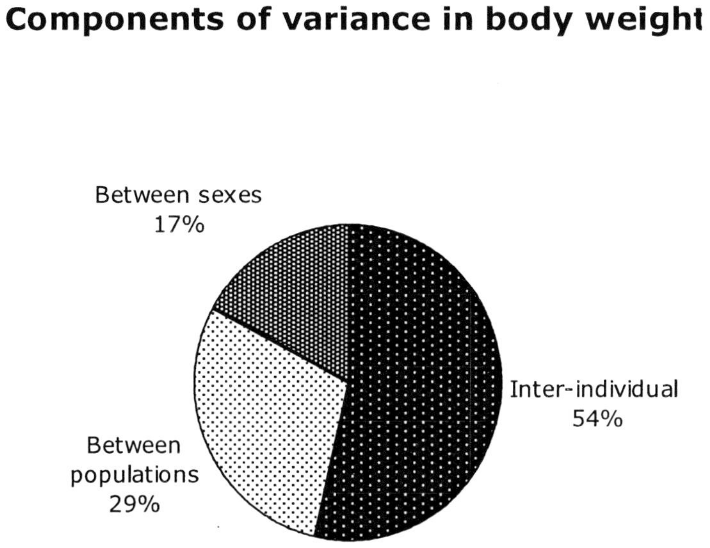

Figure 1. Components of variance in human brain size and body size (from Henneberg, 1990).

South Africa of the late 1980s and early 1990s was a land of official apartheid policies that were already recognised as being wrong, but their strong effects, creating a sharply divided society, were still visible. At liberal universities, an understanding of the biological basis for the fallacy of racial classification was welcomed and students willingly accepted an anthropological view of the nature of human variation. After the abandonment of Apartheid in 1992, as the process of integration began, social divisions were mostly of a historical socio-economic nature. Black, Indian, Cape Coloured and White people for generations had lived side by side in the same country and interacted frequently. Therefore, at an individual-to-individual level, there was a natural understanding of the lack of any "racial" differences, but the policies necessary to reverse the negative effects of apartheid–the affirmative action, still required racial re-classification of citizens, this time to improve lives of some. Nevertheless, race remained a significant concept in the life of the country. Curiously, because distinctions between Black Africans, European Africans, Indians and Coloured people were sharp, fresh and politically-based, there was little that biological anthropologists needed to do to explain them, beyond forensic identification of skeletal remains.

In Australia, by contrast, with its vast majority of European settlers, a distinct minority of Aboriginal Australians, and noticeable population of immigrants from Asia, the popular concept of race is the one based on outdated Blumenbachian concept of "Caucasoid" dominance. The term "Caucasoid" has been corrupted into "Caucasian" and is used, for political correctness, as a replacement for "British". Immigrants from Mediterranean countries, all recognised as belonging to the "Caucasoid" race in the classic scheme of Blumenbach and Coon, in Australian society are denied the membership in the "Caucasian" race and are described as "Mediterranean" or "Near Eastern". C. Loring Brace (2005) gives a good explanation of the social superiority attached to the term "Caucasian". As direct descendants of the First Parents from the Paradise on the southern slopes of Caucasus Mountains, Caucasians did not degenerate, as did Blacks and Asians who moved away from the region where the Paradise was located.

In their everyday life Australians profess fairness and equality. They will genuinely apply it to individuals, but in their assessment of group characteristics, they rely on stereotypes. This alienates fresh immigrants, and in urban areas leads to creation of ghettoes, subcultures and gangs. As in the United States, university students in Australia politely learn the explanations of human variation taught by biological anthropologists, while in practice they still rely on popular culture. The extent of this reliance is such that the same students who provide correct answers regarding the lack of human racial differentiation in final biological anthropology examinations, will unquestioningly use terms "Caucasian" and "Asian" to describe subjects of their own, or other people's research when writing papers for other courses.

CONCLUSION

Teaching about human variation in higher level university courses is ineffective in penetrating public consciousness. What is needed to change this situation is the introduction of concepts of continuous human biological variation into primary and secondary school curricula and, through the media, into the popular culture. We are remiss in not doing so a long time ago.

REFERENCES

Brace, C.L. (2005). *"Race" Is a Four-Letter Word*. New York: Oxford University Press.

Czekanowski, J. (1965). *Człowiek w Czasie i Przestrzeni*. Warsaw: PWN.

Henneberg, M. (1990). Brain size/body weight variability in *Homo sapiens*: consequences for interpreting hominid evolution, *Homo* 39:121-130.

Henneberg, M. (1997). Human evolution today - which way next? *Perspectives in Human Biology* 3: 1-12.

Henneberg, M. (2006). The rate of human morphological microevolution and taxonomic diversity of hominids. *Studies in Historical Anthropology* 4: 49-59.

Solomon, B., Ruehli, F.J., Lee, Y.C., & Henneberg, M. (2009). Secular trend in the opening of the sacral canal: an Australian study. *Spine* 34(3): 244-8.

Štrkalj, G., & Gibbon, V.E. (2007). The race concept in contemporary biological anthropology. In E.B. Bodzsar & A. Zsakai (Eds.), *New Perspectives and Problems in Anthropology* (pp.37-46). Newcastle Upon Tyne: Cambridge Scholars Publishing.

In: Teaching Human Variation
Editor: Goran Štrkalj

ISBN: 978-1-60876-616-1

Chapter 4

RACE AND GEOGRAPHIC VARIATION CONFLATED: AN IMPEDIMENT TO TEACHING HUMAN BIOLOGY

Rachel Caspari*

Department of Sociology, Anthropology and Social Work
Central Michigan University

ABSTRACT

The race concept is a taxonomic interpretation of human variation both because of the psychology of racial thinking and because of the history of the concept of race, itself. Since taxonomy and phylogeny are interrelated, a phylogenetic assumption underlies the race concept. This assumption is inconsistent with the minimal population structure within the human species. In this chapter, I focus on the conflict between assumptions embedded in the race concept and what we know about geographic variation, a conflict that impedes our understanding of human biology. Race is often conflated with geographic variation in the scientific community and among the general public so that the mere existence of variation is frequently interpreted as phylogenetic in the minds of our students, and more generally in society as a whole. This misinterpretation may have serious negative consequences. Race models

---

* rachel.caspari@cmich.edu

remain prevalent in science and society, influencing, among other things, the interpretation and treatment of health disparities among racially defined social groups. The primacy of racial thinking renders the race concept an impediment to understanding human variation. This needs to be considered and addressed when teaching human biology.

INTRODUCTION

Whenever I teach the Introduction to Biological Anthropology, the essay question on the second exam is invariably: *What is the difference between "race" and geographic variation?* A good answer involves disentangling the many meanings of race to expose the assumptions about human biological diversity that lie at the core of the race concept. The task, then, is to compare these assumptions with what we actually know about human geographic variation. From the beginning of the semester, students know what the question will be; nevertheless, it remains a very difficult question to answer. Why is the question so hard? Because race is much more than a bad idea about geographic variation; it is both a psychological and social construction, deeply embedded in our culture, which makes that bad idea about geographic variation exceptionally difficult to depose.

Race is neither simply geographic variation nor simply differences between populations, although these are often conflated with race. Because of the history of the concept of race and the psychology of racial thinking, race is a taxonomic way of viewing human variation. Race has phylogenetic assumptions because of the relationship between taxonomy and phylogeny. Because races are believed to be genealogical entities, there are implications about the biological processes that form and maintain them; geographic isolation and independent biological histories are assumptions underlying the race concept. For these reasons, the concept of race interferes with the teaching of human biological variation. When race and geographic variation are conflated, students, and even many human biologists, assume that geographic variation is phylogenetic (i.e., relates to human evolutionary development or history). In this chapter, I focus on the conflict between the assumptions embedded in the race concept and what we know about geographic variation, a conflict that impedes the understanding of human biology. Social, psychological and historical dimensions of race are discussed elsewhere (Caspari, 2003, 2009, 2010).

THREE ATTRIBUTES OF RACE: ESSENTIALISM, BIOLOGICAL DETERMINISM AND ISOLATION

The term "race" is frequently used in the biomedical literature without definition, with the implication that racially defined social categories correspond to biological populations (Gravlee, 2009). Moreover, when race is defined, but it is often defined simply as populations that differ from one another in terms of allele frequencies. Sometimes such definitions are evoked in the context of arguments for the existence of geographic races (e.g., Dobzhansky, 1944; Risch, 2002), and sometimes they are evoked by authors who are ambivalent or opposed to the race concept (e.g., Blakey, 1999; Relethford, 2009). Both of these /interpretations are prevalent in North American society in general usage; for like many scientists, the public believes that racially defined social groups are biological entities that differ from one another genetically.

There are two problems with these interpretations that are tied to deeper meanings of race, meanings derived from human psychology and from the sociopolitical history of the race concept. The first problem involves the conflation of social and biological groups. The second involves the conflation of geographic variation and race. If races were simply populations that differ from each other genetically, race and geographic variation would be equivalent and race would not impede the understanding of human biology. However, race is not simply geographic variation. Race is a taxonomic view of human geographic or populational variation that has its roots in western science.

The race concept in the context of Western science is particularly insidious because it conflates social and biological taxonomies. Writings in the 18th century natural history tradition were largely descriptive, and the Western concept of race developed within it to delineate and describe primary divisions of humankind (Marks, 1995; Brace, 2005). While there were debates within science about the number and constituents of these divisions (whether three, five or more), there was little question about what they represented: individual subspecies, or perhaps species. Race as defined by the western race concept was codified by Linnaeus in the definitive 10th edition of *Systemae Naturae*, where Linnaeus describes five subspecies of humans and lists both morphological and behavioral characteristics of each type. These traits were considered a part of the essence of racial types and were implicitly (and explicitly) understood to be part of the intrinsic biology of each race. Thus,

European prejudices were clearly incorporated into Linnaean typology (Marks, 1995; Blakey, 1999); and in *Systemae Naturae*, Africans (*Homo sapiens afer*) are described as impassive and lazy, and Europeans (*Homo sapiens europus*) as very smart. Thus, from its very inception, the Western concept of race embodied both essentialism and biological determinism, the linking of behavioral traits such as intelligence, criminality, industriousness, and other personality traits to the essence of racial categories. The categories and their essence, with both physical and behavioral components believed to be stable and unchanging, were historically defined by science as natural groups, even as they were part of a stratified social system due to the effects of colonialism and slavery. While Linnaean taxonomy applied only to continental groupings as subspecies of humans, racial taxonomies could be extended to groups lower in the taxonomic hierarchy. A major goal of anthropology in the 19th century and through much of the 20th century was to determine the exact number and types of races (Coon *et al.*, 1959). Many workers recognized a few primary races (subspecies) each with multiple subdivisions. No matter where they stood in the taxonomic hierarchy, these categories were, in effect, "biologized" social groups; the essence of these categories was presumed to be biological (Blakey, 1999).

Race remained a core concept in anthropology and human biology through much of the 20th century (Stanton, 1960; Stocking, 1968; Brace, 2005) and biological determinism represented its core value as an applied science. As a result, this racially based science reinforced the underlying assumption of the inferiority of non-dominant racial groups and provided "evidence" to support atrocious social policies from slavery to genocide (Hrdlička, 1918a, 1918b, 1918c; Gasman, 1971; Blakey, 1987; Proctor, 1988; Stein, 1988; Paul, 1998). After World War II biological determinism was ardently protested by anthropologists (e.g., Montagu, 1964; Mead *et al.*, 1968), Nevertheless, it continues to reemerge as a valid basis for study (e.g., Herrnstein & Murray 1994; Rushton & Ankey 2000). It therefore remains important to present evidence refuting biological determinism in teaching part of a curriculum addressing human variation.

Isolation, the erroneous third attribute of the race concept, has implications for how human variation is apportioned, and it is this component of the race concept that most impedes students' understanding human diversity. It also is a primary issue when the existence of race is debated in biological anthropology today (Edgar & Hunley, 2009; Caspari, 2010). Isolation can influence thinking about human variation on two levels: first, in terms of the Western race concept, isolation implies the existence of human subspecies, that is, phylogenetic (evolutionary or historical) categories that correspond to large continental geographic groups. Second, a deeper assumption associated with *racial thinking* is that relationships in human populations are also phylogenetic, and therefore that tree structures validly model relationships between human populations.

Historically, tree models are important components of race: races are depicted as clades, sharing a common ancestor, on an evolutionary tree. While race concepts are rarely deconstructed in everyday life, in teaching human variation, explanations of human variation need to be consistent with the type concept for the race concept to be valid scientifically. Trees, as a way taxonomic hierarchies are conceived and depicted, embody an explanation for the concordance of physical and behavioral traits that are believed to make up race. Historically, the use of the tree format has rendered thinking about race very similar to thinking about biological species. This tendency is exemplified in the polygenism prevalent in the American and French schools of thought that dominated much of anthropology for the first half of the 19th century (Stanton, 1960; Stocking, 1968; Brace, 1982). Even after widespread acceptance of evolution and many of the elements of Darwinian theory, a form of polygenism continued to thrive (Wolpoff & Caspari, 1997; Caspari, 2003). Taxonomic categories, including sub-species ones, continued to be conceptualized as discrete groups, while the essences of the categories were explained as products of separate evolutionary histories. Thus, races, like species categories, came to be depicted as branches on an evolutionary tree, whose differences were explained through their independent evolution, and whose inequality was explained in terms of different selection pressures and rates of evolutionary change. As discussed elsewhere (Wolpoff & Caspari, 1997), human evolutionary literature is filled with examples of evolutionary polygenism that were used to validate biological determinism (e.g., Vogt, 1864; Haeckel, 1883, 1905; Klaatsch, 1910, 1923; Hooton, 1931; Keith, 1936; Coon, 1962).

"Minor races" within racial hierarchies were also depicted as smaller branches within major ones (Coon, 1939, 1965), and while the authors cited above believed in long, separate evolutionary histories, accounting for big differences between continental racial groups, the use of short branches represented the same assumptions on a smaller scale. Branches, whatever their length, imply divergence and isolation, a pattern that does not depict relationships between human populations that are interconnected by gene flow. While tree models, in themselves, do not necessarily represent racism (for instance, short branches within humanity have been used to underscore human similarities and the "brotherhood of man," depicting a recent common ancestor of all populations), these models are completely in accord with the essentialism of the race concept and the hierarchical structure of taxonomies. Phylogenetic models are an intuitive way to biologically account for discrete essences.

It is telling that the phylogenetic assumption has largely been used, untested, throughout the history of anthropology. Because of the power and prevalence of the race concept, phylogenetic relationships among human populations have often been simply assumed. Recently, however, more complex discussion have developed concerning the types of models to best represent human genetic diversity (e.g., Hunley *et al.*, 2009).

Therefore, the biological race concept is not equivalent to human variation; it has three attributes that make it a specific way of interpreting variation: 1) essentialism, which is intrinsic to taxonomies and, which itself may have a biological basis (Atran, 1990; Hirschfeld, 1996, 1998; Gil-White, 2001; Atran *et al.* 2002; Prentice & Miller, 2007), 2) biological determinism, and 3) because of the relationship between taxonomy and phylogeny, the assumption that human population relationships are phylogenetic. Biological determinism, as a link between the concept of race and racism, has been ardently addressed by anthropologists since the 1960s, partially in reaction to expressions of racism within the scientific community and their use in political dialog and action (e.g., Gates, 1944; Garrett, 1961; Coon, 1962; George, 1962; Putnam, 1962). However, the concept of race continues to reemerge, partially because the other two attributes of the race concept are so difficult to dispel: that is, essentialism and the assumption that human population relationships are phylogenetic are deeply entrenched aspects of racial thinking.

THE STUDY OF HUMAN VARIATION: UNDERMINING RACIAL ASSUMPTIONS

All three attributes of race concepts are undermined by the study of the biology of human variation. While a focus on intraspecific evolutionary processes, and therefore gene flow between populations, caused scientists of the modern synthesis to question the validity of subspecific taxonomic categories (Mayr, 1942; Wilson & Brown, 1953), the assumptions of race were questioned much earlier. In American anthropology, Franz Boas challenged many dimensions of the race concept. As early as 1894 he explicitly rejected racial determinism of culture:

> Historical events appear to have been much more potent in leading races to civilization than their faculty, and it follows that achievements of races do not warrant us to assume that one race is more highly gifted than another (Boas, 1894a: 303).

Boas also published against the stability of types and emphasized the complexity of human variation, challenging the essentialism at the root of the race concept (e.g., Boas, 1894b, 1911, 1912, 1918). He was interested in the lack of concordance in the geographic distribution of different features and by 1918 he had mapped clines of different traits in an attempt to understand human variation (Boas, 1918). Boas was also interested in the apportionment of human diversity, which directly addresses the taxonomic assumptions underlying race. Presaging Lewontin (1972) by more than half a century, Boas wrote in *The Mind of Primitive Man* (Boas, 1911: 94) "...differences between different types of man are, on the whole, small as compared to the range of variation within each type." Therefore, many elements of the arguments against race that emerged in the 1960s and 1970s (Brace, 1964a, 1964b; Livingstone, 1962; Lewontin, 1972) were present long before the modern synthesis.

The modern synthesis, however, brought focus to intraspecific evolutionary processes that challenged the validity of subspecific taxonomies. Studies of clines were introduced and the term "population" replaced the term "race" as description of the focus of study. Because of the nonconcordance of genetically independent features, clines came to be seen as an alternative way to study human variation that focused on individual features, rather than races or populations. Promoted by Livingstone (1962), the study of clines to describe the distribution of individual morphological and genetic traits came to replace race as a focus of analysis for many workers. Livingstone argued against using subspecific taxa, not only because they are arbitrary and difficult to define, but also because the phylogenetic assumption inherent in taxonomy obscures the complex causes of their variation: migration and geneflow represent intraspecific evolutionary processes which are overlooked in phylogenetic models.

Other authors focus on the term "population" as an alternative to "race" (e.g., Brues, 1972). This usage addresses the importance of local groups that many thought to have more biological integrity than the larger geographic delineation of "races" (Wilson & Brown, 1953). However, these discussions do not indicate that "populational thinking," the antithesis to essentialism (Mayr, 1982), had been adopted (Caspari, 2003). The concept of populational thinking emphasizes all the processes that reduce or increase variation within and between populations. Gene flow is critical to populational thinking; and its action combined with that of other evolutionary forces explains the distribution of variation within species. Gene flow undermines racial thinking. Mayr himself has suggested that the concept of populational thinking that he promoted helped to cause the demise of the race concept, as essentialism is the antithesis to Darwinian approaches to variation.

However, populational thinking has not necessarily been applied to the study of populations. Many anthropologists and other human biologists, whether studying genes or morphology, continued to think racially: they viewed populations as isolated from other groups. Some recognized this implicitly; some explicitly, as Garn wrote (1962: 6) "the contemporary approach to race stems from population genetics, where a race is viewed as a breeding population, neither more nor less." He identified small "local races" like the "Bushmen" of South Africa as more or less *isolated* breeding populations (Garn, 1962). In spite of Washburn's (1963) admonition that races or populations should be considered open systems, populations were thus conceptualized as closed. Therefore, the existence of types was implicit. As Armelagos *et al.* (1982) points out, many studies in skeletal biology and genetics continued to employ typological methods to typological ends: for the recognition and delineation of populations. Their conclusion in 1982 was that whether using skeletal or genetic traits, many studies of populations are just as typological as studies of race (Armelagos *et al.*, 1982) in that populations were assumed to be phylogenetic groups.

Although many genetic studies remain typological (Brown & Armelagos, 2001; Caspari, 2003), modern population genetics has done more than any other discipline to undermine the phylogenetic assumption inherent in the race concept. Perhaps the most often cited scientific challenge to the race concept is Lewontin's 1972 paper emphasizing that very little human genetic variation is attributable to racial or even populational differences. Using Sewall Wright's "F" statistic, originally developed to assess inbreeding, Lewontin shows that an assumed hierarchical population structure explains very little of the genetic variation in humans. F_{st}, the proportion of the total variation attributable to genetic differences between groups, is very low: about 6% for regional differences and 8% at the subpopulation level. The vast majority of the variation is found within subpopulations. These low levels of interpopulational diversity are supported by a large number of more recent studies (e.g., Nei & Roychoudhuri, 1982; Barbujani *et al.*, 1997; Jorde *et al.*, 2000; Rosenberg *et al.*, 2002; Serre & Pääbo, 2004; Manica *et al.*, 2005; Prugnolle *et al.*, 2005; Barbujani & Belle, 2006). By showing minimal population structure on several levels, these works undermine the phylogenetic assumption not only on the level of subspecies, but also on level of population. This actuality has been underscored by Templeton (1998), who argues that human populations have such little structure that "treeness" is not demonstrated and phylogenetic models are invalid. A number of more recent studies also underscore the fluidity of populations, showing strong correlations between genetic variation and geographic distance (Manica *et al.*, 2005; Prugnolle *et al.*, 2005), conforming to an isolation by distance model, with few discrete boundaries between populations (Serre & Pääbo, 2004). By challenging the phylogenetic assumption, these studies refute not only the western race concept, but also racial thinking in general.

These and other recent studies underscore the complexity of the causes of human genetic diversity, complexity caused by variation in gene-flow and selection, which undermines the taxonomic assumption. Since the genetics of populations reflect population histories (including large scale migrations, population fissioning and reticulation, marriage patterns and an assortment of other social variables), and perhaps most importantly, natural selection (Hawks *et al.*, 2007; Coop *et al.*, 2009), the apportionment of diversity among human populations is extremely variable (Hunley *et al.*, 2009; Long *et al.*, 2009). The genetic data to date indicate that there is a clinal pattern for both neutral and adaptive genetic variation, and that this pattern is affected by variable population expansions and population structure. This complexity undermines the taxonomic assumptions of the race concept. Despite this, however, even among anthropologists and other students of human variation, multinational studies show that a significant number of scholars still accept the idea that biological races exist (Littlefield *et al.*, 1982; Lieberman *et al.*, 1992, 2003a, 2003b, 2004; Kaszycka & Štrkalj, 2002; Kaszycka & Strzałko, 2003a, 2003b; Kaszycka *et al.*, 2009). This acceptance may reflect the conflation of race and geographic variation.

CONCLUSION

The delineation of human variation negates many of the assumptions of race as well as the concept of race itself. Because of the complexity of human population relationships, there is no taxonomy below the species level. For some students of anthropology, the claim against race is taken to mean that there is no geographically structured biological variation, so that any genetic clustering of populations (or classically defined races) indicates that races exist (Risch *et al.*, 2002), or that it is valid to model race phylogenetically. This straw-man concept underscores the conflation of race and geographic variation. Clearly there are biological differences between human groups. This has always been clear; even the low levels of intergroup variation suggested by Lewontin are significant (Hunley *et al.*, 2009). At issue for the race concept is how significant population structure is in characterizing relationships between populations, and whether those relationships should be considered taxonomic.

In the study of human variation, because race and geographic variation are so often conflated, the existence of variation is interpreted as phylogenetic in the minds of our students, and more generally in society as a whole. This interpretation may have serious negative consequences. Race models remain prevalent in science and society, influencing, among other things, the interpretation and treatment of health disparities among racially defined social groups: medical conditions due to social inequality (Schell, 1997; Williams & Collins, 2001) are frequently treated as the consequence of different unique ancestries (Schwartz, 2001; Dressler *et al.*, 2005; Gravlee *et al.*, 2005; Frank, 2007; Gravlee, 2009). Moreover, because race and geographic variation are conflated, the recognition of the biological attributes of social categories, from health risks associated with poverty to those associated with aspects of an individual's ancestry (such as sickle cell anemia), may serve to reify race (Duster, 2005; Braun, 2006). Thus, although the study of population disparities is critical for many reasons, the conflation of race with geographic variation may make social equality even more difficult to achieve.

All these observations underscore the importance of refuting race when teaching human biology; yet it remains difficult for students to discard the concept of race. Despite a semester of study, many of my students continue to conflate race and geographic variation because the idea of race is so deeply embedded in our culture, and taxonomic thinking seems to be a part of the human condition. The fact is that the primacy of racial thinking renders the race concept an impediment to understanding human variation, which is an issue that needs to be considered and addressed by those who teach human biology.

ACKNOWLEDGEMENTS

I thank Goran Štrkalj for the invitation to participate in this volume.

REFERENCES

Armelagos, G.J., Carlson, D.S., & Van Gerven, D.P. (1982). The theoretical foundations and development of skeletal biology In F. Spencer (Ed.), *A History of American Physical Anthropology 1930-1980* (pp. 305-328). London: Academic Press.

Atran, S. (1990). *Cognitive Foundation of Natural History*. New York: Cambridge University Press.

Atran, S., Medin, D.I., & Ross, N. (2002). Thinking about biology: modular constraints on categorization and reasoning in the everyday life of Americans, Maya, and scientists. *Mind & Society* 3(2):31-63.

Barbujani, G., & Belle, E.M.S. (2006). Genomic boundaries between human populations. *Human Heredity* 61:15-21.

Barbujani, G., Magagni, A., Minch, E., & Cavalli-Sforza, L.L. (1997). An apportionment of human DNA diversity. *Proceedings of the National Academy of Sciences USA* 94: 4516-4519.

Blakey, M. (1987). Skull doctors: intrinsic social and political bias in the history of American physical anthropology, with special reference to the work of Aleš Hrdlička. *Critique of Anthropology* 7: 7-35.

Blakey, M. (1999). Scientific racism and the biological concept of race. *Literature and Psychology* 45:29-43

Boas, F. (1894a). Human faculty as determined by race. *Proceedings of the American Association for the Advancement of Science* 43: 301-327.

Boas, F. (1894b). The half-blood Indian, an anthropometric study. *Popular Science Monthly* 45: 761-769.

Boas, F. (1911). *The Mind of Primitive Man*. New York: Macmillan.

Boas, F. (1912). Changes in the bodily form of descendents of immigrants. *American Anthropologist* n.s.14: 530-562.

Boas, F. (1918). Note on the anthropology of Sweden. *American Journal of Physical Anthropology* 1: 415-426.

Brace, C.L. (1964a). A nonracial approach towards the understanding of human diversity. In M.F.A. Montagu (Ed.), *The Concept of Race* (pp. 103-152). New York: Free Press.

Brace, C.L. (1964b). On the race concept. *Current Anthropology* 3: 13-20.

Brace, C.L. (1982). The roots of the race concept in physical anthropology. In F. Spencer (Ed.) *A History of American Physical Anthropology 1930-1980* (pp. 11-29). London: Academic Press.

Brace, C.L. (2005). *Race is a Four-Letter Word: The Genesis of the Concept.* New York: Oxford University Press.

Braun, L. (2006). Reifying human difference: the debate on genetics, race and health. *International Journal of Health Services* 36: 557-573.

Brown, R.A., &, Armelagos, G.J. (2001). Apportionment of human genetic diversity: a review. *Evolutionary Anthropology* 10:34-40.

Brues, A. (1972). Models of clines and races. *American Journal of Physical Anthropology* 37: 389-399.

Caspari, R. (2003). From types to populations: a century of race, physical anthropology and the American Anthropological Association. *American Anthropologist* 105: 65-76.

Caspari, R. (2009). 1918: Three perspectives on race and human variation. *American Journal of Physical Anthropology* 139: 5-15.

Caspari, R. (2010). Deconstructing race: racial thinking, geographic variation and implications for biological anthropology. In C.S. Larsen (Ed.), *A Companion to Physical Anthropology* (pp. 104-123). New York: Wiley-Blackwell Publishing.

Caspari, R. & Wolpoff, M.H. (1996). Coon and Weidenreich. *Human Evolution* 11: 203.

Coon, C.S. (1939). *The Races of Europe*. New York: MacMillan.

Coon, C.S. (1962). *The Origin of Races*. New York: Knopf.

Coon, C.S. (1965). *The Living Races of Man*. New York: Knopf.

Coon, C.S., Garn, S.M., & Birdsell, J.B. (1950). *Races: A Study of the Problems of Race Formation in Man*. Springfield: C.C. Thomas.

Coop, G., Pickerell, J.K., Novembre, J., Kudaravalli, S., Li, J., Absher, D., Myers, R.M., Cavalli-Sforza, L.L., Feldman, M.W., & Pritchard, J.K. (2009). The role of geography in human adaptation. *PLOS Genetics* 5(6): e1000500. doi:10.1371/journal pgen.1000500

Dressler, W.W., Oths, K.S., & Gravlee, C.G. (2005). Race and ethnicity in public health research: models to explain health disparities. *Annual Review of Anthropology* 34(1): 231-252

Dobzhansky, T. (1944). On species and races of living and fossil man. *American Journal of Physical Anthropology* 2: 251-265.

Duster, T. (2005). Race and reification in science. *Science* 307: 1050-1051.

Edgar, H.J.H., & Hunley, K.L. (2009). Race reconciled?: How biological anthropologists view human variation. *American Journal of Physical Anthropology* 139: 1-4.

Frank, R. (2007). What to make of it? The (re)emergence of a biological conceptualization of race in health disparities research. *Social Science and Medicine* 64: 1977-1983.

Gasman, D. (1971). *The Scientific Origins of National Socialism: Social Darwinism in Ernst Haeckel and the German Monist League*. New York: Elsevier.

Garn, S.M. (1962). *Human Races* (revised edition). Springfield: C.C. Thomas.

Garrett, H.E. (1961). The equalitarian dogma. *Perspectives in Biology and Medicine* 4: 480-484.

Gates, R.R. (1944). Phylogeny and classification of hominids and anthropoids. *American Journal of Physical Anthropology* 2: 279-292.

George, W.C. (1962). *The Biology of the Race Problem*. Washington D.C,: National Putnam Letters Committee.

Gil-White, F. (2001). Are ethnic groups biological species to the human brain? *Current Anthropology* 42: 515-553.

Gravlee, C.C. (2009). How race becomes biology: embodiment of social inequality. *American Journal of Physical Anthropology* 139: 47-57.

Gravlee, C.C., Dressler, W.W., & Bernard, H.R. (2005). Skin color, social classification and blood pressure in southeastern Puerto Rico. *American Journal of Public Health* 95: 2191-2197.

Haeckel, E. (1883). *The History of Creation, or the Development of the Earth and its Inhabitants by Natural Causes. A Popular Exposition of the Doctrine of Evolution in General, and that of Darwin, Goethe, and Lamarck in Particular*. New York: Appleton.

Haeckel, E. (1905). *The Wonders of Life*. New York: Harper.

Hawks, J., Wang, E.T., Cochran, G.M., Harpending, H.C., & Moyzis, R.K. (2007). Recent acceleration in human adaptive evolution. *Proceedings of the National Academy of Sciences USA* 104: 20753-20758.

Herrnstein, R.J., & Murray, C. (1994). *The Bell Curve: Intelligence and Class Structure in American Life*. New York: Free Press.

Hirschfeld, L.A. (1996). *Race in the Making: Cognition, Culture and the Child's Construction of Human Kinds*. Cambridge, Mass.: MIT Press.

Hirschfeld, L.A. (1998). Natural assumptions: race, essence and taxonomies of human kinds. *Social Research* 65(2): 331-349.

Hooton, E.A. 1931. *Up From the Ape*. New York: MacMillan.

Hrdlička, A. (1918a). Physical anthropology: its scope and aims; its history and present status in America. A. Physical anthropology, its scope and aims. *American Journal of Physical Anthropology* 1: 3-23.

Hrdlička, A. (1918b). Physical anthropology: its scope and aims; its history and present status in America. B. History. *American Journal of Physical Anthropology* 1: 133- 182.

Hrdlička, A. (1918c). Physical Anthropology: its scope and aims; its history and present status in America. C. Recent history and present status of the science in North America. *American Journal of Physical Anthropology* 1: 267-304; 377-414.

Hunley, K., Healy, M.E., & Long, J.C. (2009). The global pattern of gene identity variation reveals a history of long-range migrations, bottlenecks and local mate exchange: implications for biological race. *American Journal of Physical Anthropology* 139: 35-46.

Jorde, L.B., Watkins, W., Bamshad, M., Dixon, M., Ricker, C., Seielstad, M., & Batzer. M. (2000). The distribution of human genetic diversity: a comparison of mitochondrial, autosomal, and Y-chromosome data. *American Journal of Human Genetics* 66: 979- 988.

Kaszycka, K.A., & Štrkalj, G. (2002). Anthropologists' attitudes towards the concept of race: the Polish sample. *Current Anthropology* 43: 329-335.

Kaszycka, K.A., Štrkalj, G., & Strzałko, J. (2009). Current views of European anthropologists on race: influence of educational and ideological background. *American Anthropologist* 111: 43-56.

Kaszycka, K.A., & Strzałko, J. (2003a). "Race" – still an issue for physical anthropology? Results of Polish studies seen in the light of the U.S. Findings. *American Anthropologist* 105: 116-124.

Kaszycka, K.A., & Strzałko, J. (2003b). Race: tradition and convenience or taxonomic Reality? More on the race concept in Polish anthropology. *Przegląd Antropologiczny – Anthropological Review* 66: 23-37.

Keith, A. (1936). *History from Caves: A New Theory of the Origin of Modern Races of Mankind.* London: British Speleological Association Publication.

Klaatsch, H. (1910). Menschenrassen und Menschenaffen. *Korrespondenzblatt der Deutschen Gesellschaft für Anthropologie, Ethnologie, und Urgeschichte* 41: 91-99.

Klaatsch, H. (1923). *The Evolution and Progress of Mankind.* New York: Stokes.

Lewontin, R.C. (1972). The apportionment of human diversity. *Evolutionary Biology* 6: 381– 398.

Lieberman, L., Hampton, R.E., Littlefield, A., & Hallead, G. (1992). Race in biology and anthropology: a study of college texts and professors. *Journal of Research in Science Teaching* 29(3): 301-321

Lieberman, L., Kaszycka, K.A., Martinez-Fuentes, A.J., Yablonsky, L., Kirk, R.C., Štrkalj, G., Wang, Q, & Sun, L. (2004). The race concept in six regions: variation without consensus. *Collegium Antropologicum* 28(2): 907-921.

Lieberman, L., Kirk, R.C., & Corcoran, M. (2003a). The decline of race in American physical anthropology. *Przegląd Antropologiczny-Anthropological Review* 66: 3-21.

Lieberman, L., Kirk, R.C., & Littlefield, A. (2003b). Perishing paradigm: race - 1931-1999. *American Anthropologist* 105: 110-113.

Linnaeus, C. (1758). *Systemae Naturae* (10th edition). Stockholm: Laurentii Salvii.

Littlefield, A., Lieberman, L., & Reynolds, L.T. (1982). Redefining race: the potential demise of a concept in physical anthropology. *Current Anthropology* 23: 641-655.

Livingstone, F.B. (1962). On the non-existence of human races. *Current Anthropology* 3: 279- 281.

Long, J.C., Li, J., & Healey, M.E. (2009). Human DNA sequences: more variation, less race. *American Journal of Physical Anthropology* 139: 23-34.

Manica, A., Prugnolle, F., & Balloux, F. (2005). Geography is a better determinant of human genetic differentiation than ethnicity. *Human Genetics* 118: 366-371.

Marks, J. (1995). *Human Biodiversity: Genes, Race, and History*. New York: Aldine de Gruyter.

Mayr, E. (1942). *Systematics and the Origin of Species*. New York: Columbia University Press.

Mayr, E. (1982). *The Growth of Biological Thought: Diversity, Evolution, and Inheritance*. Cambridge, Mass.: Belknap Press of Harvard University Press.

Mead, M., Dobzhansky, T., Tobach, E, & Light, R.E. (Eds.) (1968). *Science and the Concept of Race*. New York: Columbia University Press

Montagu, A. (Ed.) (1964). *The Concept of Race*. New York: Free Press.

Nei, M. &, Roychoudhury, A.K. (1982). Evolutionary relationships and the evolution of human races. *Evolutionary Biology* 14: 1-59.

Paul, D.B. (1998). *The Politics of Heredity: Essays on Eugenics*. Albany: SUNY Press.

Prentice, D.A., & Miller, D.T. (2007). Psychological essentialism of human categories. *Current Directions in Psychological Science* 16(4): 202-206

Proctor, R. (1988). From *Anthropologie* to *Rassenkunde* in the German anthropological tradition. In G.W. Stocking Jr. (Ed.), *Bones,Bodies, and Behavior: Essays on Biological Anthropology* (pp. 138-179). Madison: The University of Wisconsin Press.

Prugnolle, F., Manica, A., & Balloux, F. (2005). Geography predicts neutral genetic diversity of human populations. *Current Biology* 15: R159-R160.

Putnam, C. (1962). *Race and Reason: A Yankee View*. Washington D.C.: Public Affairs Press.

Relethford, J.H. (2002). Apportionment of global human genetic diversity based on craniometrics and skin color. *American Journal of Physical Anthropology* 73: 629- 636.

Relethford, J.H. (2009). Race and global patterns of phenotypic variation. *American Journal of Physical Anthropology* 139: 16-22.

Risch, N., Burchard, E., Ziv, E., & Tang, H. (2002). Categorizations of humans in biomedical research: genes, race and disease. *Genome Biology* 3(7): comment 2007.1-2007.12.

Rosenberg, N.A., Pritchard, J.K., Weber, J.L., Cann, H.M., Kidd, K.K., Zhivotovsky, L.A., & Feldman, M.W. (2002). Genetic structure of human populations. *Science* 298: 2381-2385.

Rushton, J.P., & Ankney, C.D. (2000). Size matters: a review and a new analysis of racial differences in cranial capacity and intelligence that refute Kamin and Omari. *Personality and Individual Differences* 29: 591-620.

Serre, D., & Pääbo, S. (2004). Evidence for gradients of human genetic diversity within and among continents. *Genome Research* 14: 1679-1685.

Schell, L.M. (1997). Culture as a stressor: a revised model of biocultural interaction. *American Journal of Physical Anthropology* 102: 67-77.

Schwartz, R.S.(2001). Racial profiling in medical research. *The New England Journal of Medicine* 344: 1392-1393.

Stanton, W. (1960). *The Leopard's Spots: Scientific Attitudes toward Race in America 1815- 59*. Chicago: University of Chicago Press.

Stein, G.J. (1988). Biological science and the roots of Nazism. *American Scientist* 76: 50-58.

Stocking, G.W. Jr. (1968). *Race, Culture, and Evolution: Essays on the History of Anthropology*. New York: Free Press.

Templeton, A.R. (1998). Human races: a genetic and evolutionary perspective. *American Anthropologist* 100: 632-650.

Vogt, C. (1864). *Lectures on Man: His Place in Creation and in the History of the Earth.* London: Longman, Green, Longman and Roberts.

Washburn, S.L. (1963). The study of race. *American Anthropologist* 65: 521-532.

Williams, D.R., & Collins, C. (2001). Racial residential segregation: a fundamental cause of racial disparities in health. *American Journal of Public Health* 93: 200-208.

Wilson, E.O., & Brown, W.L.Jr. (1953). The subspecies concept and its taxonomic implications. *Systematic Zoology* 2: 97-111.

Wolpoff, M.H., & Caspari, R. (1997). *Race and Human Evolution.* New York: Simon and Schuster.

In: Teaching Human Variation
Editor: Goran Štrkalj
ISBN: 978-1-60876-616-1

Chapter 5

AN IN-CLASS EXERCISE ON HUMAN VARIATION

Patricia C. Rice*
Division of Sociology and Anthropology,
West Virginia University, Morgantown, USA

ABSTRACT

This in-class exercise on human variation gives instructions on how instructors can introduce students to data collection and analysis in introductory level bioanthropology classes. Students collect data on themselves, acting both as a population of "natives" and as professional bioanthropologists. They then turn raw individual data into group data, and calculate phenotype, genotype, and allele frequencies on simple Mendelian traits, such as phenylthiocarbamide (PTC) tasting, or do averages and ratios on more complex traits, such as stature. The instructor can use this data base along with simulated data on four ancestral generations to help students understand microevolution, the Hardy-Weinberg Law, causes of evolution, and even "race."

* E-mail: pat.rice

INTRODUCTION

A survey of over 300 undergraduate students at West Virginia University suggests that very few college or university undergraduates are given the opportunity to gather and analyze data of any kind. This informal survey was taken on students in six of my introduction to bioanthropology classes. Students took the class for science distribution credits, so the survey included students majoring in all disciplines from biology to physical education. Sociology majors, for example, seldom collect any social data for analysis, relying instead on other people's data for insight into social problems. Political science majors seldom collect data even during election years, again relying on professionals to gather and analyze data; and likewise, biology majors seldom count fruit fly mutations, relying on work done in the 1930s. How many anthropology majors collect first-hand data in any of the four subfields, and then analyze it for "knowledge gained," whether that knowledge takes the form of lessons in methodology or in hypotheses to be tested using other data? Probably not many.

A different survey, this time on texts for biological anthropology, suggests there are three major sections of a typical introductory course in biological anthropology: (1) background material covering the basic topics of genetics, geological timetables, evolutionary biology including causes of evolution, dating methods, primate behavior; (2) human evolution from the beginning of life or first primates through the basic "grades" of human evolution to modern humans; and (3) the study of human variation today. The exercise described in this chapter focuses on the third part of a basic course. This exercise has a number of advantages: it allows students to gather data, something they rarely do; it allows them to play the role of both "natives" giving data and professionals allowed to gather it; it allows students to analyze real, raw data and turn it from data on a number of individuals into meaningful group data; it allows them to use that group data along with simulated data on ancestral generations to understand difficult concepts such as the Hardy-Weinberg Principle/Law and population genetics; it allows them to do genetic/evolution studies using only simple percentages rather than statistics, something mathematically challenged students appreciate; and finally, the students find the exercise to be fun as well as educational.

Students initially gather and analyze "real" data; subsequent data and analyses are fictionalized because real circumstances do not exist to collect and analyze data for multiple generations on a single population. The exercise works best in a class of 35-50 students. The exercise can end at the data and initial analysis stage, taking three class periods. Or, the exercise can continue up to three more class periods adding fictional generations of data to the original data to explore human variability, microevolution, population genetics, likely causes of evolution, and "race." The exercise requires minimum equipment and can be done in the regular classroom. (See Figure 1 for a list of equipment needed for each trait, as appropriate.)%%%

The exercise is detailed in the sections below, and the time requirement for each exercise is indicated.

THE CLASS EXERCISE

For Mendelian Traits:

ABO Blood: if you decide to have students do blood analysis themselves, allow enough time for this in class. You can purchase a kit from Lab Aids. It is easier to just ask students if they know their blood type or have them make one up.

PTC tasting: borrow strips from biology department or purchase PTC tasting strips. Lab Aids provides kits with 50 strips and a more detailed analysis than this exercise requires for US $17.95. Labaids.com will give latest cost and instructions. Carolina Biological sells just PTC strips for US $4.10 for 100 strips. You can cut the strips into 4 strips each. You can either purchase "control" strips of plain paper or make them yourself.

Mid Digital Hair: just observation, but be sure students are looking at the mid digits and not the digits closest to the knuckles.

Darwin's Point: just observation. You can draw a picture of the trait on the Data Sheet.

Tented Fingers: just observation.

Hitchhiker's Thumb: just observation.

Tongue Roll: just observation.

Ear Lobe Attachment: just observation. You can draw pictures of free hanging vs. attached lobes on the Data Sheet.

Short Index Finger: using a piece of cardboard slightly larger than a palm, draw the outline of one of your hands, with fingers slightly spread. Tape a small piece of cardboard at the end of the ring (4th) finger so "natives" will bump the end of that finger against the barrier. Then it is easy to see if index finger is shorter or not.

Hair Whorl Direction: just observation. Sometimes the hair at the whorl point needs to be prodded to see which direction it reveals.

Wrist Tendons: just observation, but often difficult to observe.

Color Blindness: borrow book from library, borrow from your local optometrist, or make color photocopies of 3 or 4 examples and staple them together; Ishihara Test for Color Blindness is the standard, but there is one color photo in Lewis (2007) and a different one in Mange & Mange (1999).

For Continuous Traits:

Height/stature: make measurement column by using masking tape: mark off with permanent marker by cm up to 2m, marking each 10 cm by number; roll marked tape onto a pen and then unroll and tape onto wall. Provide ruler or similar for measuring from head to wall marker.

Sitting height: make measurement column by using masking tape: mark off with permanent marker by cm up to 1m, marking each 10 cm by number; roll marked tape onto pen. Using the chair "natives" will sit on, mark on wall where sitting mark occurs and tape column to wall at that point. Provide ruler or similar for measuring from head to wall marker.

Cormic Index: if you tape the standing and sitting height Data Sheets to the wall, the student in charge of the Cormic Index trait can copy data from each sheet and then do ratios of the two measurements for each "native."

Cephalic Index: a spreading caliper is needed. They are relatively expensive, so try to borrow one from your anatomy department. Or purchase from Carolina Biological or some other scientific equipment company. The Internet should help.

Nasal Index: a sliding caliper is needed. See above.

Nasal Profile: draw pictures on the Data Sheet showing the 3 stereotyped profiles: convex, straight, concave. (Exaggerate a bit on the drawings.)

Fingerprints: inkpad is needed. You can draw an example of each type of fingerprint on the Data Sheet: arch, loop, whorl.

Hair color: just observation, but best by designated stages from darkest to lightest.

Eye color: just observation, but best by designated stages from darkest to lightest.

Hair form: just observation, but account for "natural" vs. environmental effect and best using designated stages from most curled to straight.

Eye Form: draw a picture on the Data Sheet showing the stereotyped eye forms: inner epicanthic eye fold, middle/full fold; no epicanthic fold.

Body Hair Amount: do not do observation! Student should ask each "native" about the degree of his or her body hair relative to others in the same age and sex category.

Figure 1. Equipment needs by trait.

PREPARATION

After you decide to do this exercise, you should announce on the first day of class and on your paper or online syllabus when the exercise will take place. I recommend that you *require* students to be present in class on data collection day or they will not understand the exercise. If a student must be absent, he or she may collect data on individuals at the beginning of the subsequent class period; doing so rather makes a point in favor of attendance when students are anxious to finalize their data and "attack" the "native" who was missing on the previous class day.

In constructing the exercise, first, decide which variable human traits you wish your students to collect data on and subsequently analyze. Class size is not a factor: If you choose to do all 24 traits and have 24 students, then obviously one student = one trait; if you have 48 students, then there can be teams of two per trait. If you choose to do only 10 traits, then again you can assign teams to collect the data. However, each student must play the role of a "native."

Choose and prepare a list of traits that students will then collect data on (see section below). Human variable traits number in the thousands, but you should restrict the list to some or all of the following because they are relatively easy to measure or ask questions about, are not embarrassing, and are appropriate for undergraduates. (Collecting data on such traits as male pattern balding, multiple nipples, or the Hapsburg Lip would be embarrassing or inappropriate for this age bracket, as is sickle cell anemia; tooth size measurements and ear wax collection would be difficult as well. I do not take data on skin color just because of its association with "race.") You can either have students do blood typing (ABO) or have them ask each "native" what his or her blood type is, depending on your college/university views on performing a simple, but invasive, test. Since the outbreak of HIV, I have found it easier to put this trait (i.e., HIV status) in the "ask" category; it is still data collection, though its analysis is not simple. You may choose not to have a student analyze this trait because of its complexity, or at a minimum, you can assign the trait to a student who can do advanced math (see end of Figure 4).

The Potential Traits

Human variable traits come in two forms: (I) simple Mendelian traits (or monofactorial or discrete variation traits) caused by one gene with but two (or in the case of ABO, three) alleles, one dominant and one recessive, and with little or no environmental effect; and (II) continuous (or multifactorial) traits caused by a usually unknown number of genes and with a considerable environmental effect.

(I) Simple Mendelian Traits (M traits)

At last count, the human genome contained 15,991 simple Mendelian traits; most are rare and cause "abnormalities" and often early death, such as with hemophilia, cystic fibrosis, and sickle cell anemia. The following 12 "benign" traits with their dominant and recessive symbols can be used: ABO, PTC tasting (Tt), Mid digital Hair (Mm), Darwin's Point (Dd); Tented Fingers (Ll), Hitchhiker's Thumb (Hh), Tongue Roll (Rr), Ear Lobe Attachment (Ee), Short Index Finger (Ff), Hair Whorl Direction (Ww), Number of Wrist Tendons (Xx), and Color Blindness (Cc). (Note that some symbols are traditional in the literature; others have been randomly assigned.)

Other traits were at one time thought to be simple Mendelian traits, but many believe there are multiple genes involved: presence of freckles, cleft chin, dimples. There appear to be no environmental effects with these traits, however.

(II) Continuous Variation Traits (C Traits)

Continuous (or multifactorial or complex) traits are not caused by a single gene, but likely by an unknown number of genes; in addition, most have considerable environmental effects. For data collection and analysis purposes, you can point out to the class that as long as the population is from a single econiche and (when doing the second part of the exercise) are in that same relatively unchanged niche for the length of the simulated generations in the exercise, environmental effects should not be a major problem. For accuracy, all measurements should be done in metric. The following 12 continuous traits can be used: stature, sitting height (yielding Cormic Index), cephalic index, nasal index, nasal profile, finger prints, hair color, eye color, hair form, eye form, and body hair amount. We know the "heritability estimate" for some of these traits, ranging from 0 for no genetic input to 1.0 if the trait is completely genetic: the height/stature trait estimate is .6; the fingerprint trait is .9.

Prepare the Data Sheets (Figure 2). You can choose to have the class work with all 24 traits or you can eliminate some to limit equipment costs. Prepare the sheets, one per chosen trait, giving one line for the data on each individual "native" by ID number, using the format suggested in Figure 2; cut and tape the collection and analysis instructions for each chosen trait, as shown in Figure 3 on the back of the appropriate Data Sheet. Photocopy as many M Trait Instructions for Calculating Frequencies as there are chosen M traits (Figure 4) and tape after analysis instructions. Draw pictures as appropriate: nasal profiles, epicanthic folds, hyperdexterity, ear lobes, finger print types, hair form, Darwin's Point. Give out ID numbers the day before data are to be collected and instruct students to give and collect data using only that ID number. (Keep a copy of assigned ID numbers with you during Data Collection day in case someone forgets his or her ID number.)

Name of collector/s ________________

Trait ____________________________

Instructions: Collect data by observation (or asking questions), not by genetic symbols. Data collection and analysis instructions for your trait are on reverse.

ID # Males	ID # Females
1 ________________	27 ________________
2 ________________	28 ________________
3 ________________	29 ________________
etc.	

Figure 2. Data sheet and instructions for data analysis.

Mendelian Traits

ABO (M) Every human has a blood type: A, B, or O. **Data collection**: either do blood typing or ask "native" what his or her blood type is. If students do not know their blood type, ask them to guess one. Write A, B, or O on data sheet. **Data analysis:** follow directions for ABO.

PTC (M) Every individual either tastes the chemical PTC or does not. **Data collection**: ask each "native" what he or she tastes on each of two strips. Give the plain paper strip first. Write Yes if person tastes the chemical or No if not on Data Sheet. **Data analysis**: follow directions for M below. T (tasting) is dominant; t (non tasting) is recessive.

Figure 3. Continued on next page.

Mid-digital Hair Presence (M) Every individual either has hair on the middle digit of his or her fingers, or does not. **Data collection**: look for hair on the mid digits of fingers (not first or third, but middle digit). Write Present or Absent on Data Sheet. **Data analysis**: follow directions for M below. M (presence) is dominant; m (absence) is recessive.

Darwin's Point (M) A rare trait (Darwin had one) on the inner ridge of the ear. **Data collection**: look at inner ridge of ears for a triangular "bump" about 1/3 of the way down ear. Write Present or Absent on Data Sheet. **Data analysis**:
follow directions for M below. D (absence) is dominant; d (present) is recessive.

Finger Tenting (M) Everyone can tent and intertwine their fingers, putting one thumb over the other. **Data collection**: ask each "native" to intertwine his or her two hands in a tent with one thumb over the other. Look at thumb position and write Left if left thumb is over right and Right if right thumb is over left. **Data analysis**: follow directions for M below. L (right over left) is dominant; l (left over right) is recessive.

Hitchhiker's Thumb (or hyperdexterity) **(M)** Some people can bend either or both thumbs backward away from their other fingers. **Data collection**: ask "native" to bend his or her thumb backward and write Yes or No on Data Sheet. **Data analysis**: follow directions for M below. H (absent) is dominant; h (present) is recessive.

Tongue Roll Ability (M) Some people can roll their tongue from both sides toward the middle. **Data collection**: ask "native" to roll his or her tongue from each side to the middle. Write Yes or No on Data Sheet. **Data analysis**: follow directions for M below. R (can roll) is dominant; r (cannot roll) is recessive.

Ear Lobe Attachment (M) Everyone's ear lobes are either free hanging or are attached directly to the side of the face. **Data collection**: look at each "native's" ear lobes and write Attached or Free on Data Sheet. **Data analysis**: follow directions for M below. E (free hanging) is dominant; e (attached) is recessive.

Short Index Finger (M) Some people have shorter index fingers than ring fingers. **Data collection**: using the cardboard "palm" provided, have "native" place palm down flat, matching thumb and little finger to the cardboard palm. Write Yes if index finger is shorter than ring finger; write No if same length on Data Sheet. **Data analysis**: follow directions for M below. F (has short index finger) is dominant; f (does not have short index finger) is recessive.

Figure 3. Continued on next page.

Hair Whorl Direction (M) The hair on the crown of the head goes clockwise on some people and counterclockwise on others. **Data collection**: check the crown of the head of each "native" for direction of hair whorl and write Clockwise or Counterclockwise on Data Sheet. **Data analysis**: follow directions for M below. W (clockwise) is dominant; w (counterclockwise) is recessive.

Wrist Tendons (M) Everyone has 2 or 3 wrist tendons. **Data collection**: ask each "native" to bend his or her hand over his or her wrist, turning it so you can see if there are two or three tendons. Write 2 or 3 on Data Sheet. **Data analysis**: follow instructions for M below. X (two tendons) is dominant; x (three tendons) is recessive.

Color Blindness (M) Rarely, an individual cannot see beyond "shades of gray," but it is basically a matter of not being able to discriminate between reds and greens. **Data collection**: using the color blindness chart, see if each "native" can see color. Write Yes (can see color) or No (cannot see color) on Data Sheet. **Data analysis:** follow instructions for M below. C (blindness is absent) is dominant; c (color blind) is recessive.

Continuous Traits

Stature (height) Every individual has stature. **Data collection**: tell the "native" to stand up straight and measure height in metric. **Data analysis**: keep males and females separate and do average for each plus average for group. Note min-max range for each.

Sitting Height Every individual has sitting height. **Data collection**: tell the "natives" to sit up straight and measure sitting height in metric. **Data analysis**: separate males and females, and calculate average for each plus average for group. Note min-max range for each.

Cormic Index This is the ratio between stature and sitting height. **Data collection**: copy both stature and sitting height from Data Sheets. **Data analysis:** calculate what proportion of each individual's stature is in his or her sitting height. (Divide sitting height by stature.) Do averages for males and females separately plus group average. You will get scores between 45 and 55.

Cephalic Index This is the ratio between the head length and width. **Data collection**: measure longest part of "native's" head between eyes and back of head keeping caliper even; measure widest part of "native's" head in front view, keeping calipers even. **Data analysis**: calculate the ratio between the two, dividing width by length; you will get ratios between 70 and 85. Do average for population and note min-max ranges.

Nasal Index This is the ratio between the nose length and width.

Figure 3. Continued on next page.

Data collection: measure length of nose from between eyes to bottom of nose and width between the nostrils. **Data analysis**: calculate the ratio between the two by dividing width by length; you will get ratios between 64 and 104. Calculate average for population and note min-max ranges.

Nasal Profile Nasal profiles are convex, straight, or concave. **Data collection**: match the three stereotyped nasal profiles and write convex, straight, or concave on Data Sheet. **Data analysis**: using code for convex (1), straight (2), and concave (3), calculate % for each category; calculate an average for the population: multiply the number of natives in each category by the code number and divide by total population number.

Finger Prints Every finger has an arch, whorl, or loop on the first digit. **Data collection**: with ink pad and paper, roll fingers to get prints. **Data analysis:** count how many arches, whorls, and loops there are for this population (l0 per individual); calculate % of each. You cannot do an average of this trait.

Hair Color Everyone has hair of one color or another. **Data collection**: ask each "native" if his or her hair is "natural." If not, ask what is and write the natural color on the Data Sheet. **Data Analysis**: use 5-color, dark-light scale of DD (5), D (4), M (3), L (2), LL (1) for trait. Count how many fall into each category and calculate % of each. Calculate average for population by multiplying the number of "natives" in each category by the code number and dividing by the total population number.

Eye Color Everyone has color in their eyes. **Data collection**: ask each "native" if his or her eyes are the "natural" color. If not, ask what is and write the natural color on the Data Sheet. **Data analysis**: use 5-color scale of DD (5), D (4); M (3), L (2) LL (1) for trait. Count how many fall into each category and do % of each. Calculate average for population by multiplying the number of "natives" in each category by the code number and dividing by the total population number.

Hair Form Everyone's hair has a form. **Data collection**: ask each "native" if his or her hair form is "natural." If not, ask what is and write the natural form on the Data Sheet. **Data analysis**: use 5-form scale of pepper corn (5), frizzy (4), curly (3), wavy (2), straight (1). Count how many fall into each category and do % of each. Calculate average for population by multiplying the number of "natives" in each category by the code number and dividing by the total population number.

Eye Form Eye form varies from an epicanthic fold over the inner part of the eye, to a full fold in the middle of the eye, to no fold. **Data collection**: observe eyelids and write inner, middle, or none on Data Sheet. **Data**

Figure 3. Continued on next page.

analysis: use 3 form scale of epicanthic fold of inner (3), middle (2), none (1). Count how many fall into each category and do % of each. Calculate average for population by multiplying the number of "natives" in each category by the code number and dividing by the total population number.

Body Hair Amount Everyone has a certain amount of hair on his or her body. **Data collection**: ask "native" if he or she has less, the same, or more body hair than others in his or her age and sex category. **Data Analysis**: keeping males and females separate, use 3 amount-scale of More (3), Same (2), Less (1). Count how many fall into each category and do % of each; do same for total population. Calculate averages for all three populations by multiplying the number of "natives" in each category by the code number and dividing by the total population number.

(Instructor: Cut and paste the instructions for only the trait assigned to each student/team to each Data Sheet, Figure 2. Photocopy and add M (Figure 4) instructions as appropriate for each Mendelian trait)

(M) = Mendelian Trait; (C) = Continuous Trait

Figure 3. Instructions for collecting/analyzing data.

Example	**Instruction**	**Fill in your data**
Phenotype frequencies (how many taste versus not taste)		
Tasters N=30 = .60 Non-tasters N=20 = .40	Count phenotypes in each category and calculate %.	Dominant N= =. Recessive N= =.
Allele frequencies (how many dominant and how many recessive alleles)		
T = .37 t = .63	Recessive is the square root of recessive phenotype type frequency. Dominant is what is left from 1.0.	Dominant = . Recessive =.
Genotype frequencies (how many individuals fall into the three categories of homozygous dominant, heterozygous dominant, homozygous recessive)		
TT = .13 **Tt = .47** **tt = .40**	Recessive is known from phenotype. To calculate homozygous dominant, square frequency of dominant allele. To calculate heterozygous dominant, add the two numbers and subtract from 1.0.	Homozygous dom = . Heterozygous dom = . Homozygous recessive = .

Figure 4. Mendelian trait (M trait) instructions for calculating phenotype, estimated genotype, and estimated allele frequencies (except for blood typing).

Have the equipment ready to use for each chosen trait. Decide ahead of time where in the classroom each trait's data will be collected. Finger prints need a hard surface; stature needs a wall for the measuring tape, and the location for measuring for standing and sitting height traits should be near each other; cardboard "palm print" needs to be taped on a flat surface, etc. Walls, tables, and the front row of desks are all suitable for taping down Data Sheets.

It is assumed that the following terms and concepts have already been covered in class; normally this is done in previous sessions, either on human genetics or on causes of evolution: genes, alleles, phenotype, genotype, gene frequency, morphology, dominant, recessive, Law of Independent Assortment, mutations, selection, gene flow, genetic drift.

Day One: Data Collection (50 Minutes)

On day one, go to class early and tape down the prepared Data Sheets; if you don't tape the sheets down, they will get lost. Place equipment at the proper data collecting "station." Scatter a dozen pens or pencils near the Data Sheets. It is helpful to have your graduate assistant or just an appropriate member of the class come in 10 minutes early to help you. Keep the class outside until you have everything set up. Write the traits on the board along with each student's identification number.

On this day, students will collect data on themselves as if under field conditions. Announce that they will be a population of say 50 individuals living on an isolated island, so each of them is a giver of data as well as a professional visiting biological anthropologist who has permission to collect simple, non-invasive and non-embarrassing data on individuals in the population. Give the population a name just for reference purposes. (For reasons given below, I always call the class/population "Population 5.")

Then start the class by distributing the prepared Data Sheets (Figure 2) to the students who are assigned individual traits. The sheets contain vital information for the Data Collection phase: the collector's name/s and trait, the identification numbers that must be used for reporting all data (to keep data as anonymous as possible), and instructions for collecting and analyzing the particular trait. Point out where each trait's "station" is around the room. Next, ask each student to stand next to his or her Data Sheet and ask them to quickly read the instructions for data collection for their assigned trait. Then tell them to begin collecting and giving data: if there are teams of two students for each trait, suggest that one team member stand by the Data Sheet and collect data as instructed, while the other team members "revolve" around the room giving his or her traits (by ID number only) to other data collectors; then when about half of the data are collected, the team members change roles. Tell students to write only what they observe on the Data Sheets at this stage. Your job is just to see that things keep moving along. Since you are the only one with the names and ID numbers of "natives," collectors will start asking you "who is number 14?" and you need to be able to point to the appropriate person.

To make sure each "native" gives all traits before leaving the classroom and to make sure he or she has taken all data, tell the class that when they believe they have both given data on each trait and taken data from each "native," they should check the trait list on the board and if the tasks are complete, then to cross off his or her number. Once all data have been collected on the Data Sheets and students have left, collect the sheets and go over them before the next class period. Make sure each Data Sheet entry is filled in properly with morphological trait observation (i.e., yes, no, 1.6m, etc.)

DAY TWO: TURNING RAW, INDIVIDUAL DATA INTO MEANINGFUL GROUP DATA – 50 MINUTES

On day two, you should bring 3 or 4 calculators to class or ask students bring their calculators.

If there were any absentees on the regular data collection day, identify who was missing. Pass out the Data Sheets to the appropriate students, having members of teams (if appropriate) sit together. Take the equipment used on the previous data collection day and have these students collect data as quickly as possible on any new "natives." Then you can announce that data collection is closed and dramatically write N=50 (or whatever your final number of "natives" is) on the board.

M trait: (using PTC as example)

PTC tasting

Phenotypes: tasters =
nontasters =
Genotypes: TT=
Tt =
Tt =
Allele frequencies:
T =
t =

C trait: (using stature as example)

Stature

Male average
Female average
Group average

Min/max range

Figure 5. Format for reporting data.

Data analysis begins with students now knowing the number of individuals in the population and with their individual/team Data Sheets in front of them. Individual Data Sheets contain previously prepared instructions for analyzing the assigned traits. Students should read the instructions, and turn their raw data on individuals into meaningful data on the group. For M traits, students will calculate phenotype, genotype, and allele frequencies. For C traits, students will find individual instructions on how they can turn the raw data into group data, keeping males and females separate in some cases. Once each individual or team has finished the analysis, they should bring their calculations to you to be checked. If you make a book (see below) of the data, after you check student results, you can give each student/team a half sheet of paper and have them write their results as indicated in Figure 5 and turn them in before they leave. Be sure to check for accuracy and make corrections as needed before proceeding.

DAY THREE: LOOKING AT DATA: 50 MINUTES

On day three, you can either make a small book or put the data online. I make a book with a cover titled with the name of the group (I call it Population 5). One advantage of making a Data Book is that you can draw pictures of some of the traits on appropriate pages before photocopying, such as hyperdexterity of thumbs or free hanging vs. attached ears. You can tape two half-sheets to make a full sheet and photocopy them, one per student, cutting them in half after photocopying. Then keeping the M traits together followed by the C traits, collate the sheets and staple to make the Data Books. If appropriate, tell students they will be using these books on occasion the following week (see below). Or you can do the work and put it online, asking students to print the data and bring the printout to class on Day 3. If you make Data Books, pass them out at the beginning of the class session.

Go over each trait, but obviously spend less time on subsequent traits. This is the time to reiterate what genes, alleles, dominance, recessiveness, range, averages mean as you look at each trait. If you do Cormic Index, you can have the shortest female and the tallest male stand at the front of the room and then sit together on a table and have the class look at their standing vs. sitting height. Invariably, the short female "gets taller" relatively speaking when she sits next to the male because males generally carry more of their height in their legs than females. You can have the hyperdexterous students show their hitchhiker's thumb or others show off their hairy fingers. If appropriate, comment on body hair amount by sex: males usually consider themselves to have equal or more hair than other males, but females usually consider themselves to have the same or less hair than other females. You can discuss the cultural reasons for this and use it as an example of why self-reporting is often a bad data collection technique. You can comment on human variability by populations: which populations have the highest amount of A blood; which populations are the tallest; which have the most arches in fingerprints,and so on.

This activity ends the first phase of the exercise: students have collected real data on an existing human population; they have turned their raw data into meaningful data; you have explained the context of the data relative to human variability. If you choose to end the exercise here, it has met the goals indicated. If you choose to continue the exercise, you will add to the established data base and use it to introduce microevolution, generational variability, population genetics, causes of evolution, and even "race."

CONTINUING THE EXERCISE: MICROEVOLUTION, POPULATION GENETICS, CAUSES OF EVOLUTION, AND "RACE" – UP TO 3 CLASS SESSIONS

Microevolution

There are many definitions of "evolution," but one that fits this exercise is that "evolution is a change in gene frequencies from one generation to the next." Since there is no way you can have students collect data on the ancestral generation(s) of the population whose data they have just collected, such generational data will have to be simulated. Tell students why you are simulating generational data: only if you have data on several generations, can you observe patterns of change in allele frequencies or averages, i.e., evolution. Five generations of data will show patterns; two or three will not. In this section, you want to demonstrate how we know evolution has taken place, not why. Show examples on the board of five generations of data, using allele frequencies of PTC and stature using average height in metric as examples of the two types of traits: (1) show an example of slow and continuous, i.e., directional, change over the five generations; (2) show an example of totally random numbers for the five generations; (3) show an example where there is little or no change over the generations, with the first and last generations being essentially the same set of numbers; (4) finally, show an example where there is a "glitch" in otherwise slow directional change, i.e., for three generations, the numbers are slowly and directionally changing, but in the next generation, there is a 15 point change, followed by the slow directional change again. In essence you are setting the class up for Sewall Wright's (1963) three "modes of evolutionary change": (1) Deterministic ("evolutionary pressure" or directional) (2) Stochastic ("random fluctuations"), and (3) Unique Events (Founder Effect), and of course, (4) no change/evolution. You can explain that if you had real data on those five generations, you could claim that you had evidence of evolution. This is the study of microevolution. Use numbers that will show the effect you wish to show: for slow and directional change, for example, using .14 for the first generation, followed by .16 for the second, followed by .19 for the third, would show the correct pattern.

Causes of Microevolution and Population Genetics

When you begin this section, tell the class that you will be using their data on one generation (group) and the simulated data on four ancestral generations to explore the causes of microevolution and the nature of population genetics. Although population genetics involves a sophisticated knowledge of statistics and the ability to predict changes in populations, what you will be doing is far simpler, though the methodology and findings are similar and instructive. Plan a lecture on the Hardy-Weinberg Law, explaining the Law and its usefulness for study of evolution. Write the H-W Principle and the hypothetical causes of evolutionary change for any population between any two (or more) generations on the board: mutations, natural selection, gene flow, genetic drift, non-random mating, unequal numbers of males and females.

Put just two generations of data (again using Tt and stature average) on the board and discuss how any one of the causes – or any combination of them -- could be the actual cause of the change. To simplify analysis, remind students that the five generations lived on an isolated island and tell them there was random mating and equal numbers of males and females, thus eliminating three possible causes of change.

Now go one step further, combining Sewall Wright's three "modes" of evolutionary change and five generations of data to assess the possible causes of evolution within the groups. Remind students of Mendel's Law of Independent Assortment (except for genes on the same chromosomes and why individual traits must be evaluated separately).

To do this part of the exercise, make up four generations of data, with the Data Book allele and averages representing the final generation, that show each of the "modes" of evolutionary change and one that shows no change but rather genetic equilibrium. You can either distribute this information as a handout or put it on the board. I leave the line blank for generation 5 so that students can fill in their own data from the Data Book. Look at the 5th generation data before you make up the ancestral four generations' data to get the correct pattern and appropriate numbers. Ask the students which mode of change each example falls into and then discuss possible causes. Only the Deterministic mode has multiple cause possibilities: mutations, selection, gene flow, genetic drift. But, recurring mutations would likely not show much change and gene flow is eliminated by the island situation. This leaves natural selection as the most likely cause of change in this mode. Ask students to speculate on possible selection factors. You can tell the class at this point that the population number averaged about 500 for each generation, not 50, so that small size and genetic drift may or may not be a factor in generational change.

This part of the exercise is extremely simplistic, but it allows students to go from looking at data on their now familiar generation to adding simulated generational data in order to assess the causes of evolution. Keep reminding them that we have no such five generational data as that would involve over 100 years of data collection on a single population. This part of the exercise also will allow you to comment on macroevolution, though of course not all agree that "macroevolution is just microevolution writ large." You can also assign a reading on real population genetics research to show what specialists really do (see e.g., Ferrell *et al.*, 1978; Gajdusek *et al.*, 1978; Gockumen, *et al.*, 2008; Zhou *et al.*, 2008).

"Race"

You can use the data on the class to discuss "race" with or without the simulated generations. If "race" is defined as "populations that have more genes in common than they have with other populations in the species," then you can assess whether the class/population is composed of different "races." Using the original Data Sheets, you can pick one trait – say PTC tasting – and compare the findings with those for another trait – say blood type -- by individual "natives": does the variability of traits correlate, trait by trait? Do "natives" with A or B blood taste PTC while "natives" with O blood do not taste PTC? Add stature. Are the A/B tasters all tall and the O non-tasters all short? By the time you add one more trait, the message should be clear: the variability of human traits does not co-vary and there is no such thing as a biological "race."

References

Ferrell, R.E., Nunez, N., Bertin, T., Labarthe, D.R., & Schull, W.J. (1978). The blacks of Panama: their genetic diversity as assessed by fifteen inherited biochemical systems. *American Journal of Physical Anthropology* 48: 269-276.

Gajdusek, D.C., Leyshon, W.C, Kirk, R.L., Blake, N.M., Keats, B., & McDermid, E.M. (1978). Genetic differentiation among populations in

Western New Guinea. *American Journal of Physical Anthropology* 48: 47-64.

Gokcumen, O., Dulik, M.C., Pai, A.A., Zhadanov, S.I., Rubinstein, S., Osipova, L.P., Andreenkov, O.V., Tabikhanova, L.E., Gubina, M.A., Labuda, D., & Schurr T.G. (2008). Genetic variation in the enigmatic Altai Kazakhs of south central Russia: insights into Turkic population history. *American Journal of Physical Anthropology* 136: 278-293.

Lewis, R. (2007). *Human Genetics* (8th edition). New York: McGraw Hill.

Mange E, J, & Mange A.P. (1999). *Basic Human Genetics* (2nd edition). Sunderland MA: Sinauer Associates.

Wright, S. (1963). Adaptation and selection. In G.I. Jepsen, E. Mayr, & G.C. Simpson (Eds.), *Genetics, Paleoanthropology, and Evolution* (365-389). New York: Atheneum.

Zhou, R., Yang, D., Zhang, H., Yu, W., An, L., Wang, X., Li, H., Xu, J., & Xie, X. (2008). Origin and evolution of two Yugur sub-clans in Northwest China: a case study in paternal genetic landscape. *Annals of Human Biology* 35: 198-211.

In: Teaching Human Variation
Editor: Goran Štrkalj

ISBN: 978-1-60876-616-1

Chapter 6

CHALLENGING UNIVERSITY STUDENTS' CONCEPTS ABOUT RACE

***Pamela Ashmore*[*] *and Donna Hart*[†]**
Department of Anthropology, University of Missouri, St Louis, USA

ABSTRACT

Anthropologists and biologists agree that the concept of biological race cannot be applied to the modern human species, but this fact is not readily available to students in higher education. Conventional ideas and the history of physical anthropology contribute to a lingering belief that the human species can be divided up into groups representative of race and that a number of characteristics are discretely (i.e., in a mutually exclusive way) distributed in each group. Modern science, however, demonstrates that many characteristics are distributed in humankind along a continuum and that typically more variation exists within a traditional racial grouping than exists between them. This chapter documents the development and implementation of inquiry-based university courses in human variation. As anthropologists teaching at an urban university, we determined that instructing students about the science of human variation is of paramount importance. We adopted an inquiry-based learning model for the courses and devised small group

[*] E-mail: ashmorep@umsl.edu
[†] hartd@umsl.edu

laboratory exercises to give students hands-on experience in collecting and analyzing data. Laboratory exercises ranged from investigating the human fossil record to determining the Mendelian genetic make up and measuring the body shape of fellow students. Typically the data do not sort into traditional racial categories. Students are introduced to the process of science and to the existence of the clinal distribution of characteristics historically considered to represent discrete groupings of humans. A pre- and post-course assessment tool is devised, and students ($n=116$) demonstrate significant changes in content knowledge (T-TEST $p<0.01$). Individual questions are tested to see if changes in historical race thinking occur. Pre- and post-assessment results document significant shifts in how students conceptualize human variation ($p<0.01$). Informing students about the science of human variation is a first step in facilitating a critical evaluation of how they view human variation in a variety of contexts.

INTRODUCTION

The concept of racial categorization is alive and well in the 21st century; in fact, we might even say it is flourishing. Barack Hussein Obama was recently elected as the 44th President of the United States of America and this was a historical moment – especially regarding the "racial" identification of the new president. Headline news from around the globe recognized Barack Obama as the first African American to be elected to the highest office in the government of the United States. President Obama is the son of a white woman from Kansas and a black man from Kenya. In A More Perfect Union which he delivered on March 18, 2008, in Philadelphia, President Obama describes having relatives of "every race and every hue." While President Obama is biracial, the tendency to label him as African American harks back to the rule of hypodescent, an American social convention in which racial identify is based on the notion that any degree of African ancestry ascribes membership to that minority race (Graves, 2001). This "rule" is based not on any known biological principle of inheritance, but reflects the social convention put in place during the 1800s when Southern states enacted "black" codes to protect the purity of the "white" race. Hypodescent, or the so-called "one drop rule," is reminiscent of the fifteenth century Spanish Inquisition, when the concept of purity of blood was enacted to prevent Jewish incursions into the rights and privileges of the dominant Christian ruling class. Jews were restricted from obtaining higher education, they could not hold political positions, and whatever property, businesses, or other economic resources they acquired were confiscated (Kamen, 1965). It is this confusion between biological facts and socio-cultural conventions that persists and continues to impact how we conceptualize race.

RACIAL CLASSIFICATION

Humans demonstrate a propensity to classify and categorize variation – including peoples' personal appearance. The ability to categorize a wide range of phenomena is an adaptive behavior, developed during the course of human evolution as it has enabled us to distinguish between friend and foe, what is safe to eat and what might be poisonous, and what is a good vs. poor quality resource for producing stone tools. This ability contributes to our survival (Brinband *et al.*, 2006) and develops in early infancy along with other cognitive abilities, such as memory, reasoning, problem solving, and learning (Batty & Taylor, 2002).

A natural inclination to categorize was demonstrated by early natural historians, anatomists, and physical anthropologists as they investigated ways to categorize human variation. In fact, the history of physical anthropology is fraught with attempts to concretely establish and document the existence of discretely different human races. In 1795 Johann F. Blumenbach, often referred to as the father of physical anthropology, provided a modern classification of human races (Wolpoff & Caspari, 1997). Based on geography and physical characteristics, and notably primarily by skull shape, he recognized five major "varieties" of humans: Caucasian, Mongolian, Native American, Ethiopian, and Malaysian. Blumenbach identified Caucasians, the European race, as the original, ideal race from which all the others derived. To his credit, Blumenbach recognized that gradations do occur between the major groups, but still he considered these major races as representing essential types reflecting the cumulative effects of different climates acting on different populations. Blumenbach's categorization of human variation was an extension of the four race classification put forth by Carolus Linnaeus in *Systema Naturae* in 1758. In this taxonomic treatise, humans are classified on the basis of color, temperament, and body positioning, or type. The resulting categories were: Americanus (red, choleric, upright); Europeus (white, sanguine, muscular); Asiaticus (pale-yellow, melancholy, stiff); and Afer or African (black, phlegmatic, relaxed) (Gould, 1981). The consequence, but not necessarily the original intent of such racial schemes, was the promotion of ideas pertaining to an inherent ranking of the races.

Samuel George Morton, a physician and natural scientist, promoted craniometric investigations by physical anthropologists (Gould, 1981). The underlying assumption in the early to mid-1800's was that a large skull size reflected a large brain size and consequently, greater intelligence. It was thought that if scientists quantified the shape and size of human skulls collected from around the world, they would ultimately yield evidence of different races. Working in Philadelphia, Morton amassed a collection of over a thousand crania providing data for physical anthropologists to scientifically *prove* the existence of human races and also to yield substantive evidence for an inherent ranking of races.

In 1859 Paul Broca established the *Société d'Anthropologie de Paris*, and for this reason is recognized as the founder of French anthropology (Wolpoff & Caspari, 1997). He greatly influenced the growth and development of the subdisciplines of physical anthropology as well as biology and medicine. His neuroanatomical research revealed that various brain functions are located in different regions of the brain, including Broca's area, the region involved in the production and formulation of spoken language. In an attempt to document a ranked order of human races, Broca amassed craniometric and morphometric data, which he believed would depict a linear progression of mental superiority. To address the primary essence of different human races, Broca also investigated hybridity. He concluded that different human races could successfully interbreed but did so with various degrees of success. He posited that the individuals of so-called hybrid races were inferior to those of pure race and more closely related races exhibited greater interfertility than other more distantly related races (Wolpoff & Caspari, 1997). Broca also established many of the methodologies and laboratory techniques still used by physical anthropologists today. He was, however, a racist, and Gould (1981) coherently explains problems inherent in Broca's treatment of the data.

Establishing the tradition of American physical anthropology, Earnest A. Hooton taught at Harvard University for 43 years (1913-1954) (Wolpoff & Caspari, 1997). Identifying race by statistical means and focusing on skeletal biology became a mainstay of his research and of the work of many of his Ph.D. candidates. Hooton distinguished between primary and secondary races and identified links between different human races and fossil anthropoids. Hooton treated races as well-established typological categories that virtually represented different species, but he resisted assigning a ranking order to the various types.

Hooton's perspective on human variation was perpetuated by his student, Carlton Coon, who also taught at Harvard and then at the University of Pennsylvania. Considering the adaptive significance of physical characteristics associated with race and adding a long-term time component to them, Coon produced a typological treatise, *The Origin of Races*, in 1962. He posited that modern humans could be divided into five races or "subspecies" which had been geographically isolated from one another for extended periods of time. He hypothesized that each subspecies independently crossed a threshold value at different points in time during their evolutionary progression to modern *Homo sapiens* (Brace, 2005). The length of time it took individual subspecies to reach this evolutionary pinnacle was reflected in the cultural achievement of each subspecies, i.e., race. The most culturally progressive Caucasoids and Mongoloids crossed first, followed by the Africans, Congoids and Capoids, and last by the southern Australoids.

By contrast, shortly after the Holocaust and World War II, physical anthropologist Ashley Montagu, sought to move away from a typological and static view of human variation and to recognize that human "races" reflect a combination of cultural and biological phenomena. In 1949, the United Nations Educational, Scientific, and Cultural Organization (UNESCO) invited an international panel of anthropologists and sociologists to compose a statement on "race" that would reflect the latest scientific knowledge. Montagu was a member of this panel and his influence is apparent as the document advocates substituting "ethnic group" for the term "race" and introduces the idea that race is not a biological reality, but instead a "social myth" (Montagu, 1951:17).

Description of the human genome confirms that the biological concept of race - that there are variants within a species that exhibit isolated biological differences, simply cannot be applied to the human species (Montagu, 1942, 1964, 1997; Lewontin, 1995; Graves, 2001; Alland, 2002; Marks, 2002). Now, in the 21st century, many physical anthropologists like us contest well established ideas and general assumptions about the existence and validity of human biological races. Furthermore, we challenge the notion that physical characteristics are discretely distributed between such groupings or that they can be used to identify distinctive racial boundaries. But ideas and concepts of race are well engrained throughout the Western world and especially so in American society.

Extended individual life histories contribute to the sustainability of such thinking. For example, it is easier to dispel a middle school student's belief in the existence of human races than to change the mindset of an adult about this concept (Ashmore & Lewis-Harris, 2007). Individual beliefs about human diversity are internalized and stored in long-term memory (Nespor, 1987; Schechtman, 1994; Senge, 1994). Such beliefs are often highly resistant to change; the more an individual has invested in a particular belief the harder it is to change (Brand & Glasson, 2004). The validity of the information on which a belief is based may not be questioned and, like a myth, it can become a reality in the mind of the individual believer (Gaston, 1970). Consequently, when young children are exposed to ideas about humans and biological race regularly associated, either covertly or overtly, with a set of values, such beliefs will become stronger over time.

The development of racial thinking in children has been documented by numerous social scientists (Lasker, 1929; Minard & University of Iowa, 1931; Horowitz, 1939; Goodman, 1946; Spencer, 1982; Katz, 1987; Ramsey & Myers, 1990). Recently a version of an Implicit Associate Test was devised for children and taken by three to five year olds. To the surprise of the project researchers, three year old children demonstrate bias for certain racial preferences and this bias is as strong in the children as in adults (Banaji, 2008). This finding underscores the existence of unconscious associations between the idea different human groups (i.e., so-called races) and various positive and negative values.

Traditional race thinking is promoted by cultural and historical assumptions that human races exist and by the belief that distinctive traits not exhibited by other such groupings can readily identify membership in a particular race.

We propose that race-based thinking will not be significantly challenged until we appropriately inform people about the reality of human biological variation and dispel the layers of myth, misconception and misperception. Like the history of racism, particularly as it unfolded in the United States attests, race-based thinking can be a powerful tool when one group uses it to undermine and diminish another group (Hutchinson, 2005). The premise behind such thinking is that racial groups constitute different sorts of human beings, with one representing some degree of superiority over another other. We suggest that in order to combat this type of thinking and behavior, we first must level the field by providing knowledge that invalidates the idea that humans can be classified and ranked using different racial categories. Race *is* a social construct, but one that is not based on scientific knowledge about biological human variation.

RATIONALE AND DEMAND FOR COURSE

As anthropologists working at the University of Missouri St. Louis, a land-grant urban research university, we were struck by the fact that a college course dealing with this subject was absent from the curriculum. Our university serves the state's largest metropolitan region that has been designated as one of the most highly segregated communities in the United States (Focus St. Louis, 2001). UM-St. Louis enrolls approximately 16,000 students, eighty four percent of these students are from the metropolitan region, and the average age of all of our students is 27. Consequently, these students represent an integral component of the work and social communities of the St. Louis region. In addition, prior to 2002, none of the seven community colleges or eight large universities in the Greater Metropolitan St. Louis region regularly offered such a course. Consequently, we felt it was particularly important to develop and institute a curriculum on the biology of human variation.

In the 2002 academic year, we launched a team-taught seminar entitled "Human Diversity and Concepts of Race," for honors college students at the University of Missouri-St. Louis. The student demand for a seat in this course was overwhelming, which greatly encouraged us. Honors courses are taught in a seminar format and are usually limited to 15 students. However, because of student interest, we decided to expand the class to accommodate 36 students, and this decision was supported by the Dean of the Honors College. After the initial semester, the course was reconfigured into two versions, one for the Department of Anthropology and one for the Honors College, and taught annually. The anthropology course is open to non-majors, who usually constitute about 75% of each class, and class size is limited to 40 students. There is consistently a waiting list for entry into the honors seminar, and over 300 students have now completed the courses.

AN INQUIRY-BASED MODEL

Due to the complex nature of the subject of human diversity, we decided to structure the course around an inquiry-based learning model. Inquiry-based models have been found to be successful strategies for student-focused, active learning in the sciences and to foster student achievement (Lott 1983; Shymansky *et al.*, 1983; Wise & Okey, 1983; Shymansky, 1984; Von Secker

& Lissitz, 1999).Thus, because the course content is deep and multifaceted, an inquiry-based model was deemed ideal because it promotes experiential learning and the development of critical thinking skills (Edelson, 2001).

We have created a series of hands-on laboratory sessions so that students can involve themselves in scientific investigations in small groups, thereby engaging in the processes of science: Students are challenged to collect data, have discussions about their own findings, and critically analyze their results. After a sessions for introduction and data collection, discussion takes place during follow-up seminar sessions, allowing students to first reflect on their findings and on how laboratory work correlates with their own mental framework regarding human diversity.

A series of laboratory exercises are also scheduled during each course. In one lab session, students trace the physical evolution in fossil human ancestors allowing a broad overview of both the great diversity in hominin species and the minimal variation now found in *Homo sapiens*. Students also examine the processes of inheritance and identify the range of variation exhibited within their own class population by working with observable Mendelian traits, such as directionality in hair whorls and the presence or absence of widow's peaks, ear wax color, and mid-digital hair. Using anthropometric techniques, students take skull and post-cranial measurements (Figure 1), another means of identifying, tabulating, and plotting out the range of variation displayed by class members. Students also acquire metrics such as cephalic and nasal indices, and discover for themselves that measurements exist along a continuum rather than in neatly divided racial groupings.

In a fortunate coincidence, we are temporarily archiving a 19th century skeletal population in the Biological Anthropology Laboratory at UM-St. Louis and have extended participation to our anthropology seminar students. After we instruct students in how to appropriately handle and respect the remains and regarding techniques by which forensic anthropologists use skeletal ethnic markers to identify ancestry, students assess a set of 32 cranial and postcranial markers on a select number of individuals. Students make both qualitative and quantitative assessments about characteristics such as orbit shape, post-bregmatic depression; nasal openings, and the length and diameter of long bones. Typically,

Figure 1. Students doing anthropometry in a lab setting.

they determine that the placement of an individual into one of three traditional racial categories is a somewhat arbitrary exercise.

This exercise lends itself to discussion of racial profiling as it relates to law enforcement in Missouri and elsewhere. Racial profiling is employed when law enforcement use ethnic markers (based on perceived physical characteristics, e.g., skin color) when making a decision to stop, search or arrest a motorist. Over a three year period (2004-2006) police in Missouri stopped African Americans at a higher rate than warranted by the proportion of African Americans of driving age within the general population (Nixon, 2006). Citizen outrage over racial profiling prompted the passage of a state law requiring law enforcement to collect data about a driver's race for each vehicle stop and to submit these data to the Attorney General. Although the revised state statutes provide a degree of accountability for law enforcement personnel and make recommendations about counseling and training of personnel found to engage in race-based traffic stops, the statute still promotes misleading assumptions about racial identification. In the national arena, similar cases of racial profiling have involved the removal of people deemed to be of Middle Eastern ancestry from commercial flights in a post 9/11 world.

In another lab activity, students investigate the geographic distribution of ABO blood group antigens. Historically, this distribution has been affected by disease and epidemics. The lab experience better prepares students to be able to determine that the distribution of blood types is not connected to race.

Skin color variation of extant, indigenous human populations also is observed by students, and they are challenged to determine the adaptive significance of light and dark skin tones. Investigating the relationship of light and dark skin color to latitude, temperature, Vitamin D from UV radiation, and the effect of UV radiation on folic acid enables students to determine that both light and dark skin tones have been important to the adaptive radiation and success of the human species. To investigate human skin color, students are asked to place thirty color chip cards representing human skin tones, in a continuum of lightest to darkest. Students then must assign the membership of each shade card to one of four "race" categories (white, Asian, Native American, and black). Students determine whether or not the racial groupings are mutually exclusive or if they overlap with one another. Students may also self-identify with a particular color card and can incorporate their own self-identity into this exercise in racial identification.

A small group activity involves a critical analysis of race-based information available on the internet. We have identified a number of web sites that focus on the topic of race. The small groups are asked to determine the validity of information and data presented, and to critique the scientific content of such sites. Previously, we discovered that many of these highly sophisticated sites reflect varying degrees of very subtle to blatant pseudoscience. We think this is a timely exercise for students to complete because of the growing tendency of students to glean information from internet sources.

The overarching rationale behind our decision to use an inquiry-based model for the courses is to enable students to discover for themselves the intricacies of human diversity. Students have first-hand experience with the resistance of even a limited range of data to sort into neat, delimited packages identifiable as different human races. In this fashion, students directly encounter the lack of biological support for human racial categorization.

The downside of our choice in models is the large investment in time and energy that it has taken to develop and implement these courses. Over the years, however, we have had success with using inquiry-based techniques in our classroom teaching. Consequently, in a small number of cases we simply needed to modify and fine tune labs that we had already created. In other cases, we either used ourselves to field-test new labs or asked for student volunteers to participate in field-testing. As is true with the implementation of any new course, we have modified lab sessions over the years but we have never completely scraped a single lab. Due to the nature of the student learning process, we caution that faculty deciding to use inquiry-based learning strategies must have a solid knowledge of the content that they are presenting (Keys & Bryan, 2001). Engaged students tend to ask more questions, and they expect faculty to be involved in the process. Because of the diversity of the learners, faculty simply cannot anticipate the range of questions that may arise and must be prepared accordingly.

IMPLEMENTING THE COURSE

At the beginning of the course, we discuss the origin of racial concepts and the history of our own discipline, which has been largely responsible for institutionalizing race thinking. Gould's (1981, 1996) classic, *The Mismeasure of Man*, is a wonderful resource for faculty to glean interesting tidbits about the history of scientific racism. Students are amused by Lombroso's anatomical stigmata and Goddard's designation of an entire group of people as "morons." Such pursuits, once thought to reflect high academic standards, provide interesting parallels when we later discuss the inaccurate correlations that have been made between race and intelligence (e.g., *The Bell Curve*).

Despite an appropriate course description, it seems as if students are a little surprised that the course content is so heavily science orientated and not focused on multi cultural activities or content. This is a deliberate strategy on our part as we did not want to confound the scientific content with political, economic, religious, or other culturally based concepts. Although we applaud the efforts of multicultural strategies, we believe that they are best pursued once the biological nature of human diversity is first identified and understood. We do, however, spotlight the unique history of racism in the United States and take a global look at how other cultures around the world identify different groupings of human beings. Whenever possible, we introduce examples from the news media about racism or racial concepts in American popular culture into the course. Students welcome the opportunity to relate their learning experiences in the course to "real world" events and circumstances.

In the honors seminar lab sessions, we ask leading questions and provide opportunities for students to respond to our questions. Students frequently reflect on several key findings. At first, they are typically surprised when their own data do not approximate traditional racial categories but rather indicate the existence of a continuum between such categories. Then, they wonder, if science, in fact, does not substantiate the existence of human biological races, then why have so many for so long invested in making sure that this concept stays alive and well? The students cite numerous examples of their being asked to identify their own racial status, ranging from financial aid and college admission applications to filling out product warranty cards. Students then ask why they have never been presented with this type of curriculum before entering a college classroom. The later two issues typically pose real dilemmas to the students enrolled in these courses.

As the semester advances and students become more familiar and comfortable with each other and with us, we are often privileged to have our students share profound experiences and realizations that they have had. Reaching this level of comfort does require us to mandate at the beginning of the semester certain behavioral protocols about respect for oneself and for others. Recounted personal experiences may entail how students have been classified as a certain "type" of person, called by a certain name, or how they have similarly classified someone else. Students of mixed ancestry talk about their own struggles to self-identify and to claim group membership, while international students often provide unique insight into our American experience with race and address issues that they have encountered in other countries. All admit that at one time or another they have demonstrated and acted on the inclination to attribute (based sometimes on skin color alone) a certain behavior or ability to another person. They also realize that when they have been on the receiving end of this type of association, it is often not comfortable and not a good fit. In any given semester, these sessions are the most revealing and substantive.

In the honors college version of the course, small groups (three or four students) are required to conduct research on a topic related to racial perceptions, racial stereotyping, or current research on human variation. The students design their own research project, including their research question or hypothesis, and also identify a discrete study population; they construct an assessment tool (such as a questionnaire), collect and analyze their data, and formulate summaries and conclusions. Projects are presented to the class and faculty as PowerPoint presentations. This further engages students in action research supplementing what they are doing in class.

These student projects have been insightful, but typically require many revisions of a group's research question, hypothesis, questionnaire or other data gathering tool. If revisions are required, review by another physical anthropologist is a good strategy because we recognize that we cannot afford to miss potentially inflammatory or insulting questions or tasks that our students are going to make public in our campus community. Consequently, more than one opinion is helpful for reviewing and critiquing these student projects. We are to some degree flexible about the range of topics that students decide to research, allowing them to incorporate culturally focused issues and topics. Among other topics, student researchers have looked at how mixed heritage individuals create uniquely personal "racial" identities, how people match physical characteristics to perceptions of professional achievement and potential academic success, how males and females categorize people in dissimilar ways, and how international students perceive diversity differently than American students.

Changing Ideas And Concepts

We developed a pre- and post-course assessment tool to give to students at the initial class meeting and at the last class meeting of each new semester. We want to determine at the onset of the course if the ideas that our students have about human variation are representative of historical "race thinking." Students are asked to answer the questions as honestly as possible, and to not be concerned about their performance on the assessment since any score is not considered as part of their grade. We implement the same assessment at the end of the course to determine if the students' ideas and concepts about race have changed.

Historical "race thinking" or folk knowledge portrays members of the human species as forming concrete biological entities representative of races. In addition, it assumes that physical characteristics can be linked to behavior and IQ, and that so-called racial characteristics like hair texture, skin color, and nose shape are inherited as a unit. Race thinking often characterizes racial traits as being passed on through human blood lines, and assumes there is an implicit ranking order of such groups based on intellectual, physical, cultural, or moral characteristics (Corcos, 1997).

The results of paired pre- and post-assessments of 20 true or false and multiple choice questions are available for 116 students from seven different separate course offerings. We created questions that would gauge biological knowledge about human variation gained during the course and scored the answers as correct or incorrect based on that perspective. Questions were also geared to evaluate whether students demonstrated typical historical race thinking or folk knowledge about race. Two-tailed t-tests were used to ascertain the existence of significant differences in mean test scores before and after the course experience. Significant variance between the mean scores on the pre- and post-assessments was discovered ($p<0.01$). Figure 2 illustrates the results of the pre- and post-assessment scores for each of the 116 students. Twenty correct answers was the highest possible score. In the pre-assessment, scores range from 4-18 correct with a mean of 11.2 correct responses. The post-assessment scores range from 10-20 with a mean of 16.3 correct.

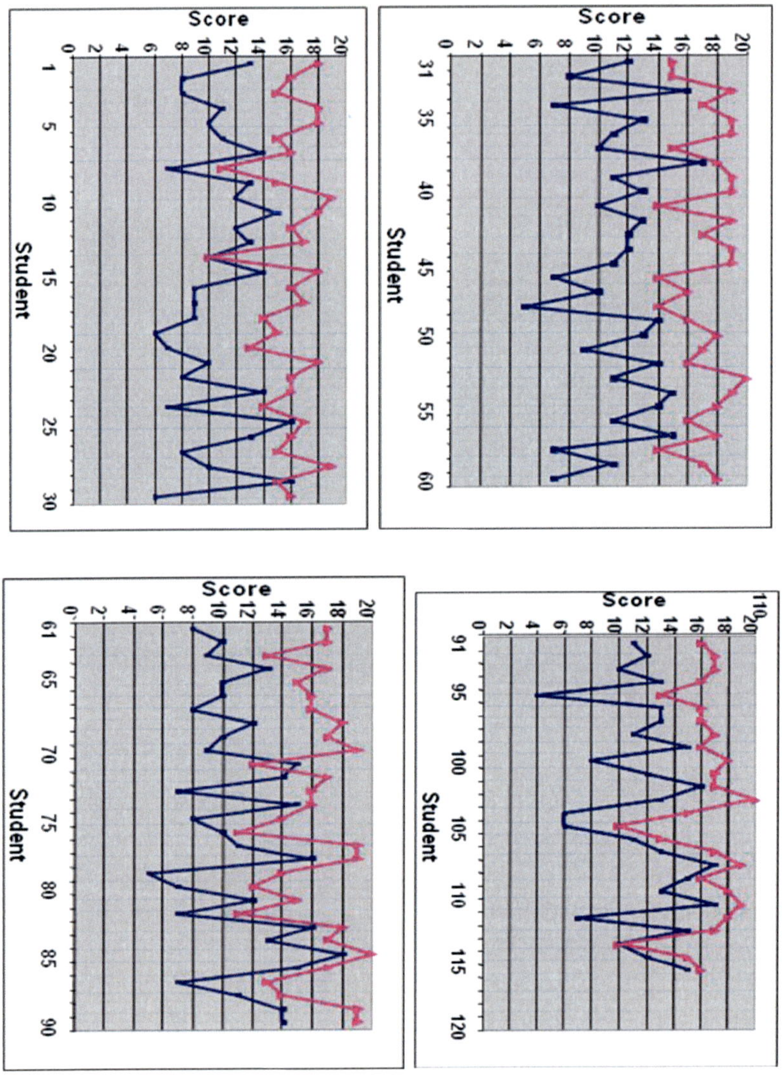

Figure 2. Comparison of pre- and post-assessment student scores . (Blue represents the pre -scores and pink the post -scores).

In our previous report on this topic (Hart & Ashmore, 2006), four questions are analyzed independently using paired t-tests. The results appear to demonstrate shifting concepts about the biological nature of human variation. We considered these questions as representative of core concepts integral to the course content. The following are the four specific questions tested:

Question 9. True or false. Human races are real biological entities.

Question 10. True or false. Genetic and physical characteristics can be used to divide the human population into different racial groups.

Question 17. Multiple choice. Race is a biological term that can be applied to a) humans, b) plants and animals, c) both a and b, d) none of the above.

Question 18. Multiple choice. Genetically, which set of characteristics is transmitted together? A) eye color, hair color, hair texture, b) skin color nose shape, lip thickness, c) height, cranial capacity, body shape, d) none of the above.

Table 1 contains the pre-and post-assessment results for the four questions. Question 9, which inquires about the biological reality of human races, demonstrates the smallest change in pre- and post mean scores (0.36). It is interesting to note that this question has the highest pre-assessment mean score (0.57). The largest change in mean scores (0.55) occurs for Question 10, regarding the use of genetic and physical characteristics to divide humans into different racial groups. The lowest pre-assessment mean scores are associated with Questions 18 and 19. Compared to Questions 9 and 10, the post-assessment mean scores are also lower. The difference between the pre- and post-mean scores for Questions 18 and 19 is the same (0.43). These questions address knowledge about the application of the biological term "race" to different species and how characteristics typically considered as indicative of racial membership are inherited.

Results of the pre- and post-course testing are illustrated in Figure 3. In order to see what the students correctly and incorrectly answered on the pre-assessment, we looked at the top and bottom distributions on the pre-test. Eighty or more students (69%) gave the correct answer to the following questions: (Question 1) ...that indigenous populations found farther from the equator have lighter skin color than those found closer to the equator, (Question 2) that people with light colored skin are more susceptible to skin cancer than people with dark colored skin, (Question 5) that sickle cell anemia is not restricted to Africans and African Americans, (Question 6) that genes do not determine whether African American males are better at playing basketball than European American males, (Question 7) that IQ is not a valid assessment of an individual's cognitive powers, and (Question 13) that "scientific racism" still exists. Students were able to identify that melanin is the term for the pigment in the body responsible for coloring (Question 14) and could select the correct geographic location of native peoples with dark skin color (Question 15).

Table 1. Four questions reflecting changing students' concepts (n=88).

Question no.	Pre-assessment mean score	Post-assessment mean score	p value	Difference in mean scores
9	0.57	0.93	<0.01	0.36
10	0.40	0.95	<0.01	0.55
17	0.10	0.53	<0.01	0.43
18	0.08	0.51	<0.01	0.43

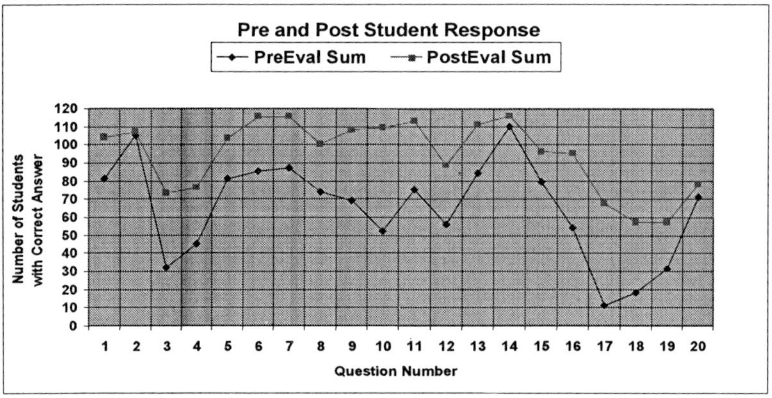

Figure 3. Pre and post-course student responses by question.

In contrast, fewer than 60 of the students (52%) responded with the correct answer to questions that addressed mechanisms of inheritance (Questions 4, 18, and 19) or human evolution (Questions 12 and 16). Students did not appear to be aware of the adaptive significance of dark skin color (Question 3), and incorrectly responded to two of the questions described above (Question 10 and Question 17). Although it is difficult to assess what the responses to these questions specifically indicate about students' concepts of race, it appears to us that questions that were more technical in content were the ones that a majority of students did not correctly answer.

CONCLUSION

Both the assessment results and our experience with these courses underscore the need to inform students about the science surrounding human variation. The results of the pre-assessment performance by students indicate that stereotypical misconceptions about human biological variation are widely held. Post-assessment findings exhibit significant change in traditional race thinking. Test results on individual questions reflecting core course content also reflect a significant change in students' knowledge about the biology of race. These findings indicate to us that the course is effective. Whether or not these changes in the way in which students conceptualize human diversity are sustainable in the long term have not, however, been investigated.

Identifying tangible and constructive strategies to indoctrinate students to the idea that there is greater diversity within so called "races" than there is between such groupings is a challenge that requires a complex mind shift. Individual and group experience, age, and the fact that the concept of human races may not have been previously challenged contribute to making this a multifaceted change in how individuals think about and perceive human variation. We feel that the opportunity to intervene or disrupt a tendency to engage in historical race thinking is worth the effort.

Students reveal that they have not previously encountered content pertaining to the biology of human diversity. These remarks support our original suspicion that this type of content has been omitted from earlier educational experiences. The findings also make us realize how important it is that institutions of higher education provide this type of science-based content to students in a wide range of disciplines.

As a result of being exposed to the course content, students now question the logic behind demands made on them to inform a variety of sources about their racial identity. Moreover, they critique the categories that are presented to them for such self-identification. As we know these categories are often insufficient, they cross cut properties such as skin color, geographic origin, and language and simply reveal the fact that such categorization of the human species is not accurate. We are generally pleased that our students leave the course struggling with these issues and thinking more critically about the topic of human diversity.

Race as a social construct cannot be merged with scientific data about human variation. Facilitating the ability of our students to recognize that these are two entirely different realms may function to offset stereotypical racial thinking. Historically, the discipline of physical anthropology greatly contributed to the perpetuation of the concept of human biological races. We feel that we now have a responsibility to challenge typological race based thinking by promoting a constructive, holistic, and science-based approach to understanding human variation. Providing students with the means to understand the science of human variation provides them with knowledge that they may then choose to use to critically evaluate economic, political, and social issues relating to human variation.

REFERENCES

Alland, A., Jr. (2002). *Race in Mind: Race, IQ, and Other Racisms*. New York: Palgrave Macmillan.

Ashmore, P.C., & Lewis-Harris, J. (2007). Addressing race through applied anthropology. *Kroeber Anthropological Society Papers*, 95: 23-39.

Batty, M., & Taylor, M. (2002). Visual categorization during childhood: an ERP study. *Psychophysiology* 39: 482-490.

Banaji, M.R. (2008). Thinking in black and white. *The Chronicle Review*, July 25, 2008, B19.

Brand, B. R., & Glasson, G. E. (2004). Crossing cultural borders into science teaching: Early life experiences, racial and ethnic identities, and beliefs about diversity. *Journal of Research in Science Teaching* 41: 119-141.

Corcos, A. F. (1997). *The Myth of Human Races*. East Lansing: Michigan State University Press.

Edelson, D. C. (2001). Learning-for-use: A framework for the design of technology- supported inquiry activities. *Journal of Research in Science Teaching* 38: 355- 385.

Focus St. Louis. (2001). *Racial Equality in the St. Louis Region: A Community Call to Action*. St. Louis: Focus St. Louis.

Gaston, P.M. (1970). *The New South Creed: A Study in Modern Mythmaking*. New York: Alfred A. Knopf.

Goodman, M. E. (1946). Evidence concerning the genesis of interracial attitudes. *American Anthropologist* 48: 624-630.

Gould, S.J. (1981). *The Mismeasure of Man*. New York: W.W. Norton & Company.

Gould, S.J. (1996). *The Mismeasure of Man: Revised and Expanded.* New York: W. W. Norton & Company.

Graves, J.L., Jr. (2001). *The Emperor's New Clothes: Biological Theories of Race at the Millennium.* New Brunswick: Rutgers University Press.

Grinband, J., Hirsch, J., & Ferrera, V. P. (2006). A neural representation of categorization uncertainty in the human brain. *Neuron* 49: 757-763.

Hart, D., & Ashmore, P. (2006). Changing students' understanding of race. *Anthropology News*47(7): 10-11.

Horowitz, R.E. (1939). Racial aspects of self-identification in nursery school children. *Journal of Psychology* 7: 91-99.

Hutchinson, J.F. (2005). *The Coexistence of Race and Racism: Can They Become Extinct Together?* Lanham: University Press of America.

Kamen, H.A.F. (1965). *The Spanish Inquisition.* London: Weidenfeld and Nicolson.

Katz, P.A. (1987). Developmental and social processes in ethnic attitudes and self- identification. In J.S. Phinney, & M.J. Rotheram (Eds.), *Children's Ethnic Socialization: Pluralism and Development* (pp. 92-99). Beverly Hills: Sage Publications.

Keys, C. W., & Bryan, L. A. (2001). Co-constructing inquiry-based science with teachers: Essential research for lasting reform. *Journal of Research in Science Teaching* 38: 631-645.

Lasker, B. (1929). *Race Attitudes in Children.* New York: Holt, Rhinehart and Winston.

Lewontin, R.C. (1995). *Human Diversity.* New York: Scientific American Library.

Lott, G.W. (1983). The effect of inquiry teaching and advance organizers upon student outcomes in science education. *Journal of Research in Science Teaching* 20: 437- 451.

Marks, J. (2002). *What It Means to Be 98% Chimpanzee: Apes, People, and Their Genes.* Berkeley: University of California Press.

Minard, R.D., & University of Iowa. (1931). *Race Attitudes of Iowa Children by Ralph D. Minard.* Iowa City: University of Iowa Studies in Character 4(2).

Montagu, A. (1942). *Man's Most Dangerous Myth: The Fallacy of Race.* New York: Columbia University Press.

Montagu, A. (1951). *Statement on Race.* New York: Henry Schuman.

Montagu, A. (Ed.). (1964). *The Concept of Race.* London: Free Press of Glencoe, Collier- Macmillan.

Montagu, A. (1997). *Man's Most Dangerous Myth: The Fallacy of Race* (6th edition). Walnut Creek: AltaMira Press.

Nespor, J. (1987). The role of beliefs in the practice of teaching. *Journal of Curriculum Studies* 19: 317-328.

Nixon, Jay (Missouri Attorney General). Executive summary on 2006 Missouri vehicle stops. 2006 [cited 2007 July 3]. Available from: http://www.ago.mo. gov/racialprofiling/2006/racialprofiling2006.html.

Ramsey, P.G., & Myers, L.C. (1990). Salience of race in young children's cognitive, affective, and behavioral responses to social environments. *Journal of Applied Behavioral Psychology* 11: 49-67.

Schechtman, Z. (1994). Challenging teacher beliefs via counseling methods: A descriptive study. *Teaching Education* 6: 49-67.

Senge, P. (1994). *The Fifth Discipline Fieldbook: Strategies and Tools for Building a Learning Organization.* New York: Doubleday.

Shymansky, J. (1984). BSCS programs: Just how effective were they? *American Biology Teacher* 46: 54-57.

Shymansky, J.A., Kyle, W.C., & Alport J.M. (1983). The effects of new science curricula on student performance. *Journal of Research in Science Teaching* 20, 387-404.

Spencer, M.B. (1982). Personal and group identity of black children: An alternative synthesis. *Genetic Psychology Monographs* 106: 59-84.

Von Secker, C.E., & Lissitz, R.W. (1999). Estimating the impact of instructional practices on student achievement in science. *Journal of Research in Science Teaching* 36: 1110-1126.

Wise K.C., & Okey, J.R. (1983). A meta-analysis of the effects of various science teaching strategies on achievement. *Journal of Research in Science Teaching* 20, 419-435.

Wolpoff, M., & Caspari, R. (1997). *Race and Human Evolution: A Fatal Attraction.* Boulder: Westview Press.

In: Teaching Human Variation
Editor: Goran Štrkalj
ISBN: 978-1-60876-616-1

Chapter 7

THE BAD OLD DAYS OF ANTHROPOLOGY REVISITED: TEACHING HUMAN VARIATION THROUGH THE PORTAL OF THE EXPERIMENTAL HISTORY OF SCIENCE

Goran Štrkalj[1,2,*] ***and Robyn Beirman***[1,†]
[1]Department of Chiropractic, Macquarie University, Sydney, Australia
[2]Institute for Human Evolution, University of the Witwatersrand, Johannesburg, South Africa

ABSTRACT

This paper argues that the history of science - specifically the experimental history of science - may be profitably used in the education of biological anthropologists for teaching human variation. E. O. Manoiloff's research in racial serology conducted in the mid-1920s is used as an example. It is suggested that in addition to reading texts on the history of human variation studies, students could repeat Manioloff's experiments using both original and modern equipment. These learning activities will demonstrate to students the weaknesses of the typological approach and racial paradigm, contribute toward the development of a

* E-mail: goran.strkalj@mq.edu.au
† rbeirman@els.mq.edu.au

better understanding of the scientific process, and cultivate an appreciation of past and present research on human variation.

INTRODUCTION

The value of history in science education has been debated for years (e.g., Debru, 1997; Allchin, 2002). It would appear that most working scientists and science educators now accept that the "history of science matters". One of the great advantages of history seems to be its potential to provide better insight into how science and particular disciplines function, especially when perceived within the broader social context. In effect, history has at least started to position itself within science curricula, including that of biological anthropology. However, the way in which the history of science is taught may be called into question (Allchin, 2002). The "science textbook history" is often presented in a simplistic, hagiographic and presentist manner, thus causing the opposite of its intended outcome, as it imparts a rather distorted, "romanticised" perception of the scientific process. It may be argued, however, that the history of science, if introduced using a more topical approach and presented in a creative way, may be of value in facilitating an understanding of the discipline to its future practitioners.

While insight into the past might be of value to all scientific disciplines, in some it seems to be almost essential. Biological anthropology is undoubtedly such a discipline, particularly its sub-field dealing with human variation. In his textbook *Human Biodiversity*, Marks (1995: 2-3) notes "The study of biological variation in the human species is thus a bit different from other kinds of scientific endeavours. Biologists studying fruit-flies certainly have the same cultural prejudice of an era and class, yet it is generally difficult to imagine those cultural prejudices pervading their work. And it is more difficult to imagine those prejudices in their work as the basis of scientific authority to oppress or to degrade lives of other people. ... Anthropology on the other hand, can and has been used in precisely those ways." Consequently, "Anthropologists ... are absorbed in their intellectual history – in learning from the mistakes of earlier generations of scholars. The more we understand those conceptual errors, which usually are visible only in hindsight, the more the science of the human species can grow – by the very process of proposal and disposal by which science functions" Marks (1995: 3). Therefore, it may be of great importance to teach students to be aware of the intellectual and ethical challenges that certain ways of thinking in anthropology (study of human variation) pose to scientists.

The study of human variation has been one of the most contentious fields of research in anthropology, if not science in general. Because of the complexity of phenomena studied, and the social implication of the research, the field has been surrounded by controversy and misunderstandings. Most past and present debate revolves around the concept of race – a concept which for a lengthy period has played a key role in describing and understanding human biological variation.

This chapter aims to demonstrate how the history of science - or to be more precise its sub-discipline of experimental history of science - could be used in teaching human variation. After dealing with the experimental history of science and human variation studies, we show how the two could be combined within a pedagogical context. Research conducted by Russian biochemist E. O. Manoiloff (sometimes spelled Manoilov) in the mid-1920s will be used as a historical example upon which the teaching activities can be based.

BACKGROUND

Experimental History of Science

The discipline of history of science abounds with different methodologies and approaches (Kragh, 1987). One of the ways in which certain historical scientific topics may be efficiently studied is through the experimental history of science. This approach is based on the assumption that "past experiments can be studied by the help of a (modern) reproduction of them" (Kragh, 1987: 160). In order to better understand the past events, one might repeat the experiments of the past using first the equipment of that day and then modern technology. One has to be careful, however, as repeating the original experiment alone will be of negligible benefit if the original theoretical framework and the social context within which science was practised at the time are not taken into account. The experimental history of science, therefore, should not be taken as the key that decodes the past, but rather as one of the useful ancillary approaches.

The "hands-on approach" of experimental history seems to be quite close to the working scientists' everyday activities, and is thus potentially usable in science education. In the field of biological anthropology (human variation studies), Gould (1978, 1981) applied methods of the experimental history of science in his well-known, provoking and controversial analysis of Morton's research on cranial capacity of different "races". Gould's results imply a series of mistakes in Morton's measurements and interpretation of data. These he attributes to Morton's racialist bias, which presumably influenced Morton's work at the subconscious level. However, subsequent re-analysis of both Morton's and Gould's research reveal serious omissions in the latter's approach, possibly based on Gould's own biases (Michael, 1988; Brace, 2005). The whole case, therefore, clearly shows the complexity of both experimental history of science and human variation studies and calls for a cautious and reflective approach (indeed, lessons could be learned from the history of the history of science).

Human Variation and "Race"

The concept of race, now generally abandoned by a majority of American and western European biological anthropologists, dominated the study of human variation for centuries (Lieberman *et al.*, 2004; Kaszycka *et al.*, 2009). Until the mid-twentieth century, the typological framework formed the foundation of research on race (the remnants of typology, it should not be forgotten, are still alive in certain research programs). Within this framework a large number of studies were generated, and as a result, racial science at the end of the nineteenth and the beginning of the twentieth century was transformed into a real classification mania that hid more than it revealed about the nature of human biological variation. Typologically based racial science came to an end only with the scientific and social changes in the second part of the twentieth century, such as the emergence of "population thinking" in biology and general social frustration with the effects of racial policies after the Second World War.

One of the ways in which population thinking reached anthropology was through the study of genetics. However, in the beginning of the twentieth century human genetics, in the form of racial serology, was figured as a different method of infraspecific classification of human species, rather than a discipline in which racialism would be questioned. Indeed, some of the early studies are paradigmatical examples of typology (Marks, 1995). One of these early studies, which today with the benefit of historical hindsight we perceive as bordering on the ridiculous if not insane, is research carried out by Manoiloff.

Manoiloff

Manoiloff seemed to be impressed with the possibility of using ABO blood typing in establishing racial boundaries. Blood, heavy with symbolic meaning (Marks, 1994, 1995), appeared to many early twentieth century scientists as a substance that contained a code that would unlock the secrets of life, thus fulfilling a role now afforded by DNA (cf. Nelkin & Lindee, 1995). It is, therefore, not surprising that Manoiloff wanted to go even further in using blood as a basis for classification. He postulated, in analogy with hormones (the presence of which in blood may be used to distinguish between the sexes), the existence of "something correspondingly specific of race in the blood of different races of mankind" (Manoiloff, 1927: 16). Indeed, Manoiloff (Manoilov, 1924, 1929) had already devised a biochemical method for sex determination in plants and animals, including humans. This method raised considerable interest and was implemented by other scientists with partial success (e.g., Falk & Lorberblatt 1927; Satina & Demerec, 1927; Abromavich & Gardner Lynn, 1930). As the same level of biological reality was ascribed to "race" as it was to "sex", Manoiloff's broadening of research to race determination came naturally.

It would appear that Manoiloff went in search of the "racial ingredient" by a trial and error method in which he used different chemical reagents. Although he could not explain what this racial ingredient was, he claimed that he managed to discover the ingredients which, when applied to blood, would detect that element by colouring blood—a kind of color-coding. The reagents used were: (1) 1% alcoholic solution of methyl-blue, (2) 1% alcoholic solution of cresyl-violet, (3) 0.5% solution of silver nitrate, (4) 40% solution of hydrochloric acid, and (5) 1% aqueous solution of potassium permanganate. These chemicals were applied to blood using the following formula: "Add to 3 cc. of unheated emulsion of red blood corpuscles (3 to 5 per cent) or directly to the coagulum of blood three to four times as much (in volume) sodium chloride (physiological solution) and mix with a glass stick so as to obtain a rather thick emulsion. Then add 1 drop of the first reagent, shake; 5 drops of second reagent, shake again; 3 drops of third reagent, shake; 1 drop of fourth reagent, and, lastly, 3 to 5 drops of the fifth reagent" (Manoiloff, 1927 :18). However, he noted that in order to get better coloration it was sometimes necessary to "add not one, but two or three drops of the fourth reagent (hydrochloric acid). The same concerns also the fifth reagent, where it may be needful to add not three, but from five to eight, and sometimes even more, drops of potassium permanganate" (Manoiloff, 1927: 18). This addition proved to be one of the keys to understanding Manoiloff's apparent success (see below).

Manoiloff used this "alchemical recipe" to distinguish between two "races", namely Russian and Jewish. He used only individuals who could be certain of their ancestry, which was determined by their knowledge of at least three "truly" Russian and Jewish ancestors from both maternal and paternal sides. On application of the formula, Manoiloff discovered that Jewish blood coloured blue or blue-green as the distinct colour of cresyl violet disappeared "entirely or almost completely" (Manoiloff, 1927: 18), while with Russian blood this did not happen, as the cresyl violet remained partly insoluble. As a result, Russian blood stayed blue-red in colour. Based on this observation, Manoiloff claimed that he could establish with 91.7% accuracy to which of the two groups the individual belonged.

Therefore, it appeared that an easy method for establishing racial belonging of an individual had been devised. The method seemed to work as Manoiloff and his collaborators subsequently successfully applied it to other groups, showing a virtual rainbow of "racial" blood colours, i.e., Esthonians and Letts – reddish-brownish, Poles reddish-greenish, Koreans – reddish-violet, Kirgis – bluish-greenish, etc. (Poliakowa, 1927). Possibly the most frightening aspect of the research was its possible application in paternal testing. Manoiloff's follower Poliakowa (1927) claimed that it might be useful in establishing paternity of children of "mixed origin" because the blood of these individuals coloured differently to that of each of the parents. For example, if the father was a Pole (reddish-greenish blood) and the mother Russian (reddish) the child's blood must, when Manoiloff's reagents are applied, colour "reddish paler".

It is too easy to dismiss this research with the benefit of historical hindsight. In the 1920s, however, the study was generally accepted with very little skepticism over its methodology and results. The very fact that it was published in the *American Journal of Physical Anthropology*, then and now one of the most respected journals in the field, tells enough. However, if Manoiloff and a few of his Russian followers were discounted, no repeated experiments reporting positive results were subsequently published. Hooton, in his 1931 classic *Up from the Ape,* seemed to be one of the first anthropologists to openly raise doubts concerning Manoiloff's research. The reason for

Hooton's scepticism, however, was the fact that Manoiloff was distinguishing between ethnic groups, not biological races. He emphasised "it is inconceivable that all nationalities, which are principally linguistic and political groups, should be racially and physiologically distinct" (Hooton, 1931: 410). Hirschfeld (1938), in one of the many anti-racist tracts that appeared in the 1930s when the first reports of Nazi racial policy started to spread, ridiculed Manoiloff and noted that anthropologists in Germany struggled unsuccessfully to repeat his experiments. Indeed, Von Verschuer (1938), one of the leading German anthropologists of the Nazi era, also noted that no one had managed to repeat Manoiloff's procedures successfully.

Manoiloff's research, an extreme example of the old racial science, seems to offer itself to historical scrutiny and provides an opportunity to learn "from the mistakes of earlier generations" through the process of "proposal and disposal by which science functions" (Marks, 1995: 3). Understanding Manoiloff and comparing and contrasting his research with the present day provides a better insight into both the scientific process and past and present research on human variation.

TEACHING STRATEGY

This learning activity is suggested as appropriate for use either in the senior year of an undergraduate course in biological anthropology, or within a postgraduate program. It may be modified according to the objectives of the specific course. Ideally, it would involve a professional historian or science historian and their students (especially if they are based in the same institution). Bioethicists, social anthropologists, philosophers and sociologists could also contribute. The involvement of these experts would add new perspectives and help students develop an appreciation of the value of interdisciplinary collaboration, which reaches beyond the boundaries of natural sciences. In fact, the whole activity could be organised as a joint project of History of Science and Anthropology departments, involving lectures and students from both departments. The activity can be undertaken as: (1) a small, one-semester project on it its own (if the course requires completion of a project) or (2) a part of the practical/laboratory classes/exercises.

Strategy 1 can be carried out by a single student or by students working in a small group. Most of the activities happen as a result of the interaction between the supervisor and the students. Although this activity has a similar or identical content to that of strategy 2, it has a different structure. It can be conducted over a longer period of time, which allows for more in-depth analysis than that allowed by strategy 2 (thus, only the strategy 2 will be described in detail with only brief remarks on the strategy 1, on points of difference). Strategy 1 was implemented as a part of the final project of a year-long Honours course in Human Biology at the School of Anatomical Sciences, University of the Witwatersrand, Johannesburg, in 2005.

Strategy 2 demands more structured activity, in order to fit within the time frame allocated for practical activities. It should ideally be carried out over three two-hour sessions. The first session should comprise discussion on racial science of the late nineteenth and early twentieth century. This discussion should be based on at least one of the classical texts dealing with history of human variation studies (or selections from them) such as Stocking's *Race, Culture, and Evolution: Essays in the History of Anthropology* (1968), Marks's *Human Biodiversity: Genes, Race and History* (1995) or Brace's *"Race" is a Four-Letter Word* (2005) and the Manoiloff paper cited above as an example of racial research of the day. Some commentaries on Manoiloff's work written at the same time should also be used (e.g., Hooton's *Up from the Ape*). A number of questions should be tackled such as: What is race typology, what are its main characteristics? What is racial serology? How does Manoiloff's research relate to the racial paradigm of the day? What is race, how was it understood at the beginning of the twentieth century and how is it understood today? Are human groups analysed by Manoiloff really biological races (according to past/present definition)? How does typological differ from population approach?

In session two, practical work is commenced. Blood should be drawn from volunteers (two 5 ml samples from each) and be preserved in different test tubes. Ideally, the blood samples should be taken from individuals belonging to populations assumed to be different races (e.g., "Caucasian", "Mongoloid", and "Negroid"). If this is not possible (due to similar ancestry of the volunteers), then different nationalities or ethnic groups should be recruited (as Manoiloff did). The study should be blind, that is, students undertaking blood analysis mustn't know the identity of the subjects from whom the samples were taken.

Two sets of blood analyses should be conducted, one set based strictly on Manoiloff's procedure, and another, in which modern equipment is used on the blood samples from the both individuals. To obtain "drops" of blood (as suggested in Manoiloff's formula) scientists of the 1920s generally used the

Pasteur pipette. Using the pipette, however, (even for an experienced person) means that the volume of the drop is rather arbitrary.

The blood portioning for the first analysis, therefore, should use the Pasteur pipette. The procedure is then repeated with modern equipment where a micropipette is used and the size of the drop is standardised. The size of the drop should be calculated as the average drop size made by a Pasteur pipette. In both cases (Manioloff's and modern), visual colour estimation through simple observation is made. In addition, digital images of all specimens are taken and later (as a "take home" activity) analysed through the Corel Photo Paint programme. The possible differences in colouration of the samples are statistically analysed by Student's T-Test.

As established by previous research (Naidoo *et al.*, 2007), the results of the application of Manoiloff's original technique would show differences in the colouring of the blood samples (albeit not the vivid colours in which he described them). However, there is no pattern to the differences, as differences are individual and they do not exist between groups. When the same technique is applied with the modern instruments, although minor individual differences are still visible, there is again no difference between populations. This conclusion will be further confirmed by the statistical analysis of the blood coloration obtained for both procedures (using old and modern equipment), which show no significant difference between populations.

In accordance with the results of the previous research, one would expect that students would conclude that Manoiloff's methodology is flawed. His recommendation was that if the original formula did not provide satisfactory coloration, additional drops of the fourth (hydrochloric acid) and fifth reagents (potassium permanganate) should be added to enable the different colorations of blood. This procedure, together with the variation in the size of the drops which was a result of using Pasteur pipettes, introduces a very high level of arbitrariness, and hence affected the validity of the experiment.

The third and final session should include another discussion in which the results of both strategies are examined (the "racial" identity of the volunteers can now be revealed). The session should start with a question: where do Manoiloff's rainbow colours come from? Previously cited literature on the history of the human variation studies should be consulted again. Some of the questions and issues raised in the first session should be re-examined in light of the results. Students should infer that fraud (conscious or unconscious) must have been at the bottom of the method because Manoiloff or technicians doing the work must have known the "racial identity" of their subjects beforehand. The students should acknowledge that the reason for the observed wide range of individual differences in the procedure using a Pasteur pipette (Manoiloff's original technique) is to be found in the arbitrary size of the drop and the application of different quantities of reagents. Broader questions on the power of racial paradigm, the relationship of the theory and experiment and observation, as well as of the influence of social/ideological determinants should be discussed. Some comparisons between past and present science should be made (being careful not to fall into the "presentist trap"), such as on the "blood mystique" (past) and "DNA mystique" (present), the status of the laboratory based science, research ethics, and so on.

This project can also assist in instructing students in issues related to ethics. As this project requires approval from an ethics committee, the student can learn the 'how, what, and why' of the ethical aspects of this specific project. Discussion of ethics in scientific and anthropological research in general can also be incorporated, including enquiries such as: Why was ethical approval not required in the first half of the twentieth century? What ethical issues are raised in the research on race at that time and on human variation today? The only limitation to this activity is time itself!

PEDAGOGY

It may be argued that this overall approach has a sound pedagogical basis. One of the basic tenets of student learning is to ensure that students are actively involved learners – not passive. Allowing students to repeat Manoiloff's experiment satisfies this goal. In addition, further facilitation of learning will occur if the assessment tasks accompanying this course are structured in a way that encourages active learning. It is well accepted by educators that student learning is strongly influenced by the assessment structure, and this very real relationship should be used to maximise the efficacy of the learning. For example, the use of class participation, or student contributions to an online discussion board (especially good for introverted or shy students), may be an appropriate method by which student involvement can be enhanced.

The project also incorporates learning in the cognitive, affective and psychomotor domains. Whether one refers to Bloom's original taxonomy of cognitive processes (recall, comprehend, application, analysis, evaluation and synthesis) (Bloom *et al.*, 1956), or the revised version by Krathwohl (2002), it is clear that the project encourages the development of the "higher order" learning processes.

The practical component of the project allows for the development of skills in the psychomotor domain (where more "advanced" students can embrace the "higher order" processes), while the very nature of the subject (race / human variation) lends the opportunity for the instructors to incorporate activities which could encourage learning in the affective domain (feelings, emotions and attitudes).

CONCLUSION

The activities described above might help students to understand the paradigm of racialism and racial typology, concepts which are on the one hand at considerable odds with views of most of the contemporary anthropologists, but which still plague some aspects of modern research. These activities may also introduce a healthy dose of scientific scepticism, especially towards the power of experimental method and measuring techniques. Through the activity, students should also learn to develop a historical distance and the ability not to judge past research according to current standards of science. They should understand that any historical episode, as well as present day science, may be understood only against the broader theoretical and social context within which it is performed. Consequently, in this activity, critical analysis, reflection and discussion dealing with the study of human variation and "race" in the past and present, as well as the experiment itself, play a crucial role.

Finally, it must also be acknowledged that the activity described here might be too expensive, too invasive, logistically too demanding or simply too time consuming for some courses and institutions. The activity should be understood, however, only as one of the many possible applications of the experimental history of science in current biological anthropology education. Repeating simpler experiments, or re-measuring or recalculating results of the past research, if carried out with scientific and historical rigour and understanding and with a critical distance, can produce equally positive outcomes.

REFERENCES

Abromavich, C.E. Jr., & Gardner Lynn, W. (1930). Sex, species, and race discrimination by Manoilov's methods. *Quarterly Review of Biology* 5: 68-78.

Allchin, D. (2002). Scientific myth-conceptions. *Science Education* 787: 329-351.

Bloom, B.S., Krathwohl, D.R., & Masia, B.B. (1956). *Taxonomy of Educational Objectives: The Classification of Educational goals (Handbook 2: Affective Domain)*. New York: Longmans.

Brace, C.L. (2005). *"Race" is a Four-Letter Word*. Oxford: Oxford University Press.

Debru, C. (1997). On the usefulness of the history of science for science education. *Notes and Records of the Royal Society of London* 51: 291-307.

Falk, K.G., & Lorberblatt, J. (1927). A chemical study of the Manoilov test for the differentiation of sexes. *British Journal of Experimental Biology* 4: 305-312.

Gould, S.J. (1978). Morton's ranking of races by cranial capacity. *Science* 200: 503-509.

Gould, S.J. (1981). *The Mismeasure of Man*. New York and London: W.W. Norton.

Hirschfeld, M. (1938). *Racism*. London: Victor Gollancz.

Hooton, E.A. (1931). *Up From the Ape*. New York: MacMillan.

Kaszycka, K.A., Štrkalj, G., & Strzałko, J. (2009). Current views of European anthropologists on race: influence of educational and ideological background. *American Anthropologist* 111: 43-56.

Kragh, H. (1987). *An Introduction to the Historiography of Science*. Cambridge: Cambridge University Press.

Krathwohl, D. (2002). A revision of Bloom's taxonomy: an overview. *Theory into Practice* 41: 212 – 218.

Lieberman, L., Kaszycka, K.A., Fuentes, A.J.M., Yablonski, L., Kirk, R.C., Štrkalj, G., Wang, Q., & Sun, L. (2004). The race concept in five regions: variations without consensus. *Collegium Antropologicum* 28: 907-921.

Marks, J. (1994). Blood will tell (won't it): a century of molecular discourse in anthropological systematics. *American Journal of Physical Anthropology* 94: 59-79.

Marks, J. (1995). *Human Biodiversity: Genes, Race, and History*. New York: Aldine de Gruyter.

Manoiloff, E.O. (1927). Discernment of human races by blood: particularly of Russians from Jews. *American Journal of Physical Anthropology* 10: 11-21.

Manoilov, E.O. (1924). Weitere Erfahrungen über meine chemische Blutreaktion zur Geschlechtsbestimmung bei Menschen, Tieren und durch Chlorphyll bei Pflanzen. *Münchener Medizinische Wochenschrift* 71: 1784-1789

Manoilov, E.O. (1929). Chemical reaction of blood for definition of sex in man and animals, and dioecious plants, *American Journal of Physical Anthropology* 13: 29-68.

Michael, J. (1988). A new look at Morton's craniological research. *Current Anthropology* 29: 349-354.

Naidoo N.P., Štrkalj, G., & Daly, T.J.M. (2007). The alchemy of human variation: race, ethnicity and Manoiloff's blood reaction. *Anthropological Review* 70: 37-43.

Nelkin, D., & Linde, M.S. (1995). *The DNA Mystique: The Gene as a Cultural Icon*. New York: W.H. Freeman.

Poliakowa, A.T. (1927). Manoiloff's 'race' reaction and its application to the determination of paternity. *American Journal of Physical Anthropology* 10: 23-29.

Satina, S., & Demerec, M. (1927). Manoilov's reaction for identification of sexes. *Science* 62: 225-226.

Stocking, G.W.Jr. (1968). *Race, Culture, and Evolution: Essays in the History of Anthropology*. New York: Free Press

Von Verschuer, O.F. (1938). Rassenbiologie der Juden. *Forschungen zur Judenfrage*, 3: 15-46.

In: Teaching Human Variation
Editor: Goran Štrkalj
ISBN: 978-1-60876-616-1

Chapter 8

SHOULD HUMAN VARIATION BE TAUGHT TO MEDICAL STUDENTS?

Tracey Wilkinson[1,*], Goran Štrkalj[2,3,†] and Muhammad A. Spocter[3,4,‡]

[1]School of Biosciences, Cardiff University, Cardiff, UK
[2]Department of Chiropractic, Macquarie University, Sydney, Australia
[3]Institute for Human Evolution, University of the Witwatersrand, Johannesburg, South Africa
[4]Department of Anthropology, George Washington University, Washington D.C., USA

ABSTRACT

The issue of human variation has recently become of great interest in medical research and practice, but numerous disagreements and misunderstandings surround this complex subject. It might be argued that these negative aspects could be reduced significantly if elements of human variation were taught to medical students. Even small changes in the curriculum, if efficiently implemented, could produce considerable gains.

* E-mail: wilkinsonat@cardiff.ac.uk
† E-mail: goran.strkalj@mq.edu.au
‡ E-mail: spocter@gwu.edu

Introduction

Human biological variation is traditionally one of the research focuses of biological anthropology. However, this complex subject is of interest to experts in many other disciplines, including the humanities, as well as social and natural sciences (cf. Ellison and Goodman, 2006; Štrkalj, 2008). The fields in which the nature and origin of modern human variation have become of great importance, and have consequently been intensely investigated, include those of medical research and practice (e.g., Metrosa, 2006; Koenig *et al.*, 2008). However, disagreement and misunderstanding of biological variation are abundant in the various medical disciplines. In this paper we suggest that many of these misunderstandings are at least partly due to lack of education on the subject (e.g., Anderson, 2008). We therefore propose that the education of medical students should include an investigation of human variation and that brief investigation into the subject would yield significant gains.

"Race" and Medicine

Many issues concerning human variation are discussed in medical discourse, the usefulness of the race concept perhaps most intensely. The claim that different "races" respond differently to the same drugs, culminating in the recent release of the first patented "ethnic drug", is an example of the importance of human variation, not only in research but also in medical practice (Hacking, 2005). The key issues relating to the debate on "race" were recently articulated by Bamshad and Olson (2003: 79) when they posed the following three questions:

> Can genetic information be used to distinguish human groups having a common heritage and to assign individuals to particular ones? Do such groups correspond well to predefined descriptions now widely used to specify race? And, more practically, does dividing people by familiar racial definitions or by genetic similarities say anything useful about how members of those groups experience disease or respond to drug treatment?

The way in which biomedical researchers deal with these questions is wide and varied. An example can be seen in an issue of a leading medical journal where two contrasting views are published back to back, highlighting the schism in opinion. The first view states that "race, at the continental level, has not been shown to provide a useful categorization of genetic information about the response to drugs, diagnosis, or causes of disease" (Cooper *et al.*, 2003: 1168), while the second interpretation concludes that "there are racial and ethnic differences in the causes, expression, and prevalence of various diseases" (Burchard *et al.*, 2003: 1174).

The reasons for the existence of different views on "race" are numerous and include both scientific and external factors. A number of studies on the status of the race concept have been carried out, some of which have delineated these factors (Littlefield *et al.*, 1982; Lieberman *et al.*, 1989; Cartmill, 1998; Kaszycka & Štrkalj, 2002; Wang *et al.*, 2003; Lieberman *et al.*, 2004; Kaszycka *et al.*, 2009). It is argued that the issue of human biological variation is too complex for consensus to be reached easily, although this might change as knowledge on various aspects of the subject improves. It has also been shown that external factors, such as the broader social context, the political climate, and the way in which the scientific community is organised, play a significant role.

Bearing this in mind, it would be too optimistic to expect a quick and simple resolution of the current controversies concerning human variation (cf. Weiss & Fullerton, 2005; Marks, 2007). However, this does not mean that attempts to address these questions should not be made. It is possible that some of the controversies and misconceptions concerning race in biomedical research may be due to a lack of adequate information on human variation in medical curricula, among other things. For example, surveys on anatomists' attitudes towards the concept of race have shown, at least in South Africa, that not only is there little agreement on race, but also that serious misunderstandings of human variation are rife in South African anatomical research (Morris, 2003; Štrkalj *et al.*, 2004). The most common mistakes are confusing socio-political traits and groups with biological traits and groups, and using small populations to generalise about much larger groups not represented by the sample (Morris, 2003). Evaluation of the same surveys leads to the conclusion that anatomists' knowledge in this field is far from ideal. Since most of those surveyed were involved in the education of biomedical science students, it is likely that their ideas will be passed on to the next generation.

HUMAN VARIATION AND MEDICAL STUDENTS

Although it is difficult to generalise, it can safely be stated that the curriculum of today's medical student does not seem to address the issue of ''race'' and human variation to a significant degree. Most medical textbooks say very little about human biological variation and if they do say anything, they tend to treat this topic in a rather simplistic and outdated manner. Textbooks on human anatomy, a discipline that would be expected to deal with the issues of variation seriously, usually miss the opportunity to tackle the subject. Van de Graaff's popular *Human Anatomy* (2002), for example, does address human variation, but only briefly and crudely in one figure. The textbook presents "principal races of humans," namely the Mongoloid, Caucasoid, Negroid, people of the Indian subcontinent, Capoid and Australoid. This classification (and terminology) is reminiscent of that proposed by Coon in his infamous, long since outdated *The Origin of Races* (1962), to which, curiously, a new "race" (i.e., "people of the Indian subcontinent") has been added.

The implementation of curricular changes, however, presents a real challenge. Due to dramatic developments in many medical and related disciplines, medical school curricula endure significant change, with new subjects constantly being introduced. Consequently, in the long, intense and expensive process of educating future medical doctors, time is essential, and any additions are carefully scrutinised. To make things even more complicated, curricular changes are not uniformly implemented in all countries, or even in different universities in the same country. There is consequently a significant heterogeneity in the way medical courses are taught.

While a number of different objections might be raised against inclusion of teaching human variation in the medical curriculum, the principal one is likely to be lack of time. To this might be added a lack of appropriately qualified faculty. Indeed, recent surveys (Nesse & Schiffman, 2003; Downie, 2004) show that these are two common objections against teaching Darwinian Medicine (understanding medical conditions in the context of evolutionary history and adaptation of *Homo sapiens*), even if it is considered. While there are compelling arguments to support teaching of some elements of evolutionary theory to medical students, Darwinian medicine still remains largely under-represented in the medical curriculum.

It would appear, however, that both human variation and Darwinian Medicine could be introduced easily with good planning. If human biology is taken as one of the pre-clinical courses, tackling the question of human variation and micro-evolutionary processes would be easy and could be introduced in several lectures and practical sessions. Integrating this knowledge into real-life medical situations is more challenging, but not impossible, especially in the context of a problem-based curriculum. A number of case studies, including some well-known cases (see below), lend themselves to discussions on human biological variation and its complexity, including its medical implications. These would be best understood if studied in conjunction with elements of Darwinian Medicine. Therefore, inclusion of principles of human variation would go hand in hand with the inclusion of principles of evolutionary biology/Darwinian Medicine.

The relationship between human variation and health and disease is a complex issue. In what follows we only touch upon, in a rather simplified manner, some of the well-known examples. The distribution of certain diseases and conditions sometimes follow continental or smaller geographical patterns of humanity's dispersal, and in some cases this might appear to correspond with one of the traditional racial classifications. However, in the large majority of cases, susceptibility to disease is distributed in a different, more complicated way. Significantly, in most instances such distribution is a result of various evolutionary forces, which may interact in different ways (cf. e.g., Spocter & Štrkalj, 2007; Trevathan *et al.*, 2008). Thus, these forces produce complex patterns of distribution. For example, the same or similar selective agents may cause convergence and similar adaptations in widely separated geographical regions. Sickle cell anaemia (produced by a homozygous recessive genotype) was or still is prevalent in several distinct geographical regions: Central and East Africa, parts of the Mediterranean and the Indian subcontinent (Relethford, 2005). These regions also share a high prevalence of malaria, an infectious disease which is the major agent selecting for the increase of the sickle cell allele. The heterozygous condition (which does not cause sickle cell anaemia) provides increased resistance to malaria. The trait of lactase persistence (i.e., the ability to digest lactose and therefore milk in adulthood), on the other hand, is globally distributed in a different manner. It is present in some human populations and is most likely related to cultural practice; it represents an adaptation to dairy farming and the availability of milk throughout the year in some but not all populations who rely on this type of subsistence (many traditional dairy farmers, of course, are lactose intolerant; they rely on cultural adaptations to process and break down lactose *before* ingestion, e.g., by making various dairy products, such as cheese). The ability to digest milk is prevalent in Europeans, but rather low in Africans who often 'suffer' from lactose intolerance. However, in some African populations, such as the Fulani (traditional dairy farmers), the frequency of individuals who have lactase persistence is close to that of Europeans (Molnar, 1998).

It could be argued that at least three different types of factors, which have been widely discussed in recent years, deserve to be included in the medical curriculum:

1. Geographical or cultural isolation

   In some cases, diseases are found almost exclusively with a high prevalence in one, usually geographically or culturally isolated, group of people. Because of its isolation, endogamous reproduction within the group prevents influx of external genetic influences while perpetuating the limited gene pool of the founders (Mak et al., 2006). This may inadvertently lead to descendants bearing disease-associated genes, which is particularly evident with recessive traits. For example, Tay-Sachs disease (homozygous recessive) is found in approximately 1 in every 2,500 births in Ashkenazi Jews compared to 1 in 500,000 births in the rest of the world (Molnar, 1998). These high rates of incidence might be due to the complex interplay between genetic drift and a strong founders' effect (Stearns, 1999), but the role of natural selection in perpetuating this disease still remains a plausible explanation with tuberculosis being one possible selective agent (Marks, 1995). Archival data suggests that pre-20th century Askenazi Jews in Europe had lower tuberculosis rates than other populations living in the same areas and that ancestors of families with Tay-Sachs disease showed increased resistance to tuberculosis (Myrianthopoulos & Aronson, 1966). Similar examples of rare Mendelian disorders with links to a particular, culturally isolated group are known to exist in the Amish, Samaritan and French Canadian populations, for example (Burchard et al., 2003).

2. Multiple interactions between genes and the environment

   Complex genetic disorders such as some forms of cancer and diabetes are likely to be a result of multiple interactions between genes and the environment (Burchard et al., 2003). These genetic associations remain for the most part poorly understood, but a few examples demonstrating the populational associations in gene frequency have been identified. For example, susceptibility to Crohn's disease is associated with three polymorphic genetic variants in the

CARD 15 gene occurring in people of European origin (Hugot et al., 2001). None of the three genetic variants were found in a similar Japanese sample also presenting with Crohn's disease (Yamazaki et al., 2002). This suggests that the disease in the group of European origin may be directly linked to the association of the CARD 15 variants, whereas other factors, perhaps ecological in nature, may be at work in the Japanese population. Another example, of particular importance in sub-Saharan Africa and Asia, is the CCR5-delta 32 variant, which offers carriers some protection against the Human Immunodeficiency Virus (HIV) by blocking its entry into cells through the CCR5 receptor (Mak et al., 2006). While this variant is virtually absent in non-European population groups (Stephens et al., 1998), as many as 25% of people of European extraction (especially Northern Europeans) are heterozygous carriers, suggesting populational differences in HIV protection.

3. Responses to drugs and drug metabolism

   Pharmacogenetics is a discipline that aims to identify genetic variants that influence responses to drug and drug metabolism. For example, genetic variants of N-acetyltransferase 2 (an enzyme involved in negating the harmful affects of certain carcinogens and the metabolism of some commonly used drugs) result in two phenotypes, slow and rapid acetylators. Slow acetylation phenotypes range from approximately 14% among East Asians, 34% among African Americans and 54% among people of European origin (Yu et al., 1994). These genetic variants are important because they predict the toxic effects of drugs and may contribute to population variation in the incidence of environmentally induced cancers.

It should not be forgotten that what appears to be a genetically determined susceptibility to certain diseases might be the result of cultural and/or ecological factors, rather than biological differences. Special attention should therefore be paid to conditions caused by cultural practices (Kuru among Fore Melanesians), changes in ecology (Japanese encephalitis in East and Southeast Asia, the Indian subcontinent and some of the Pacific Rim countries) and the complex interplay of the two conditions (including megaloblastic anaemia among South Asians) (Mak *et al.*, 2006).

These examples demonstrate the complex distribution of some ailments and indicate the potential benefit of a biologically informed knowledge of human variation. Armed with this knowledge, clinicians would be able to make better patient evaluations by familiarizing themselves with the history of the communities they serve and extending the quality of care to these communities (Betancourt *et al.*, 2003). By considering the history of a population, we may also focus attention on cross-cultural issues in health and aid in addressing racial and ethnic disparities (Betancourt *et al.*, 2003).

That understanding biological variation is of value to the clinician is apparent from the above examples, but its effectiveness will depend on its correct usage and application. Human variation is still misused in clinical practice, and certain scholarly articles bear testament to this (e.g., Garcia, 2004; Braun *et al.*, 2007). It is only through addressing the subject that we can resolve these types of misunderstanding and show when and how such knowledge may be useful in the clinical setting.

Understanding the evolutionary history of an affected population and the complex interplay of its cultural and biological factors would undoubtedly benefit medical students. However, would these relatively small additions to the medical curriculum be enough to make a real impact on students' understanding of human variation? It has been demonstrated that even a modicum of education can have a strong impact on student attitudes towards race (Štrkalj & Wilkinson, 2005; Hart & Ashmore, 2006, this volume). In studies where student opinions were elicited before and after short modules on human variation, it was observed that not only did their knowledge of human variation improve, but there was also a statistically significant shift towards discarding the traditional race concept (Štrkalj & Wilkinson, 2005; Hart & Ashmore, 2006, this volume). It might be assumed that both the course material and the views of the lecturer play an important role in this process. These examples show that even a small amount of teaching can dramatically improve students understanding of the subject and can change their (often simplistic) views on human variation. If student opinion can be changed significantly in this way, perhaps future medical researchers, practitioners and educators might be expected to have a better understanding of the subject after similar exposure to the subject. Medical educators may then influence the next generation that they teach.

Finally, the lack of qualified teaching staff among the medical faculty need not be an insurmountable difficulty. Many medical schools employ biological anthropologists in their pre-clinical departments. If not, a link could be forged with an anthropology department. With their experience and knowledge of the subject, biological anthropologists would appear to be best qualified to carry out teaching of human variation (rather than, for instance, geneticists, social scientists or the experts from humanities [Anderson, 2008]). Indeed, the focus of this type of instruction should be on microevolutionary processes underlying human variation and how these processes relate to health issues. Biological anthropologists experienced in research and teaching of human variation could also present a brief course or a workshop, or a series of workshops, on human variation for the medical faculty.

CONCLUSION

It appears that the inclusion of a few lectures and practical sessions on human variation that are up to date with current developments in the field could be profitably implemented in the medical curriculum. This, of course, is a task which entails many challenges. However, it appears that the potential difficulties are outweighed by possible benefits. It is unlikely that adopting this type of curricular change would solve all the problems related to the issue of human variation in medicine, but it might be a step in the right direction.

REFERENCES

Anderson, W. (2008). Teaching 'race' at medical school: social scientists on the margin. *Social Studies of Science* 38: 785-800.

Bamshad, M.J., & Olson, S.E. (2003). Does race exist? *Scientific American* 289: 78-85.

Betancourt, J.R., Green, A.R., Carillo, J.E., & Ananeh-Firempong, I.O. (2003). Defining cultural competence: A practical framework for addressing racial/ethnic disparities in health and health care. *Public Health Report* 118: 293-302

Braun, L., Fausto-Sterling, A., Fullwiley, D., Hammonds, E.M., Nelson, A., Quivers, W., Reverby, S.M., & Shields, A.E. (2007). Racial categories in medical practice: How useful are they? *PLOS Medicine* 4: 1423-1428.

Burchard, G.E., Ziv, E., Coyle, N., Gomez, S.L., Tang, H., Karter, A.J., Mountain, J.L., Pérez-Stable, E.J., Sheppard, D., & Risch, N. (2003). The importance of race and ethnic background in biomedical research and clinical practice. *New England Journal of Medicine*, 348: 1170 – 1175.

Cartmill, M. (1998). The status of the race concept in Physical Anthropology. *American Anthropologist* 100: 651-660.

Coon, C.S. (1962). *The Origin of Races*. New York: Alfred M. Knopf.

Cooper, R.S., Kaufman, J.S., & Ward, R. (2003). Race and genomics. *New England Journal of Medicine* 348: 1166-1170.

Downie, J.R. (2004). Evolution in health and disease: the role of evolutionary biology in medical curriculum. *Bioscience Education E-Journal* 4: 3.

Ellison, G.T.H., & Goodman, A.H. (Eds.) (2006). *The Nature of Human Difference: Science, Society and Human Biology*. London: CRC Press.

Garcia, R.S. (2004). The misuse of race in medical diagnosis. *Pediatrics* 113: 1394-1395.

Hacking, I. (2005). Why race still matters. *Daedalus* Winter: 102-116.

Hart, D. & Ashmore, P. 2006. Changing students' understanding of race. *Anthropology News* 47: 10-11.

Hugot, J.P., Chamaillard, M., Zouali, H., Lesage, S., Cezard, J.P., Belaiche, J., Almer, S., Tysk, C., O'Morain, C.A., Gassull, M., Binder, V., Finkel, Y., Cortot, A., Modigliani, R., Laurent-Puig, P., Gower-Rousseau, C., Macry, J., Colombel, J.F., Sahbatou, M., & Thomas, G. (2001). Association of NOD2 leucine-rich repeat variants with susceptibility to Crohn's disease. *Nature* 411: 599-603.

Kaszycka, K.A., & Štrkalj, G. (2002). Anthropologists' attitudes towards the concept of race: the Polish sample. *Current Anthropology* 43: 329-335.

Kaszycka, K.A., Štrkalj, G., & Strzałko, J. (2009). Current views of European anthropologists on race: influence of educational and ideological background. *American Anthropologist* 111: 43-56.

Koenig B.A., Lee S.S.-L., & Richardson, S.S. (Eds.) (2008). *Revisiting Race in a Genomic Age*. New Brunswick: Rutgers University Press.

Littlefield, A., Lieberman, L., & Reynolds, L.T. (1982). Redefining race: the potential demise of a concept in physical anthropology. *Current Anthropology* 23: 641- 655.

Lieberman, L., Kaszycka, K.A., Fuentes, A.J.M., Yablonski, L., Kirk, R.C., Štrkalj, G., Wang, Q., & Sun, L. (2004). The race concept in five regions: variations without consensus. *Collegium Antropologicum* 28: 907-921.

Lieberman, L., Stevenson, B.W., & Reynolds, L.T. (1989). Race and anthropology: a core concept without consensus. *Anthropology and Education Quarterly* 20: 67- 73.

Mak, W., Cheung, R.T.F., & Ho, S.L. (2006). Biological basis of the racial disparities in health and disease: an evolutionary perspective. In E.V. Metrosa (Ed.), *Racial and Ethnic Disparities in Health and Health Care* (pp.73-100). New York: Nova Science Publishers.

Marks, J. (2007). Long shadow of Linnaeus's human taxonomy. *Nature* 447: 28.

Marks, J. (1995). *Human Biodiversity: Genes, Race, and History*. New York: Aldine de Gruyter.

Metrosa, E.V. (Ed.) (2006). *Racial and Ethnic Disparities in Health and Health Care*. New York: Nova Science Publishers.

Molnar, S. (1998). *Human Variation: Races, Types, and Ethnic Groups*. Upper Saddle River: Prentice-Hall.

Morris, A.G. (2003). Using racial terms in anatomical research. *Plexus*, 2: 9-12.

Myrianthopoulos, N.C., & Aronson, S.M. (1966). Population dynamics of Tay- Sachs Disease. I. Reproductive fitness and selection. *American Journal of Human Genetics* 18 (4): 313-327.

Nesse, R.M., & Schiffman, J.D. (2003). Evolutionary biology in medical school curriculum. *BioScience* 53: 585-587.

Relethford, J.H. (2005). *The Human Species: An Introduction to Biological Anthropology*. New York: McGraw Hill.

Spocter, M.A., & Štrkalj, G. (2007). Darwinian medicine: an evolutionary perspective on health and disease. *South African Medical Journal* 97: 144-146.

Stearns, S.C. (1999). Introducing evolutionary thinking. In S.C. Stearns (Ed.), *Evolution in Health and Disease* (pp. 3-15). Oxford: Oxford University Press.

Stephens, J.C., Reich, D.E., Goldstein, D.B., Shin, H.D., Smith, M.W., Carrington, M., Winkler, C., Huttley, G.A., Allikmets, R., Schriml, L., Gerrard, B., Malasky, M., Ramos, M.D., Morlot, S., Tzetis, M., Oddoux, C., di Giovine, F.S., Nasioulas, G., Chandler, D., Aseev, M., Hanson, M., Kalaydjieva, L., Glavac, D., Gasparini, P., Kanavakis, E., Claustres, M., Kambouris, M., Ostrer, H., Duff, G., Baranov, V., Sibul, H., Metspalu, A., Goldman, D., Martin, N., Duffy, D., Schmidtke, J., Estivill, X., O'Brien, S.J., & Dean M. (1998). Dating the origin of the CCR5-Delta32 AIDS-resistance allele by the coalescence of haplotypes. *American Journal of Human Genetics* 62: 1507-1515.

Štrkalj, G. (2008). *Beyond Race: A Prolegomenon for the Study of Human Biological Variation*. Saarbrücken: VDM Verlag.

Štrkalj, G., Ramsey, S., & Wilkinson, A.T. (2004). Anatomists' attitudes towards the concept of race. *South African Medical Journal* 94: 90-91.

Štrkalj, G., & Wilkinson, A.T. (2005). The influence of a modicum of education on students' attitudes towards the concept of race. In G. Štrkalj,

N.P. Pather and B. Kramer (Eds.), *Voyages in Science: Essays by South African Anatomists in Honour of Phillip V. Tobias's Eightieth Birthday* (pp. 179-186). Pretoria: Content Solutions.

Templeton, A.R. (1998). Human races: a genetic and evolutionary perspective. *American Anthropologist* 100: 632-650.

Trevathan, W.R., Smith, E.O., & McKenna, J.J. (Eds.) (2008). *Evolutionary Medicine and Health: New Perspectives.* New York: Oxford University Press.

Van de Graaff, K.M. (2002). *Human Anatomy* (6th edition). New York: McGraw-Hill.

Wang, Q., Štrkalj, G., & Sun, L. (2003). On the concept of race in Chinese biological anthropology: alive and well. *Current Anthropology* 44: 403.

Weiss, K.M., & Fullerton, S.M. (2005). Racing around, getting nowhere. *Evolutionary Anthropology* 14: 165-169.

Yamazaki, K., Takazoe, M., Tanaka, T., Kazumori, T., & Nakamura, Y. 2002. Absence of mutation in the NOD2/CARD15 gene among 483 Japanese patients with Crohn's disease. *Journal of Human Genetics* 47: 469-72.

Yu, M.C., Skipper, P.L., Taghizadeh, K., Tannenbaum, S.R., Chan, K.K., Henderson, B.E., & Ross, R.K.1994. Acetylator phenotype, aminobiphenyl-hemoglobin adduct levels, and bladder cancer risk in white, black, and Asian men in Los Angeles, California. *Journal of the National Cancer Institute* 86: 712-6.

In: Teaching Human Variation
Editor: Goran Štrkalj
ISBN: 978-1-60876-616-1

Chapter 9

HUMAN VARIATION: HOW TO COUNTERACT THE NEW WAVES OF RACISM?

Charles Susanne[*]
Free University of Brussels, Belgium

ABSTRACT

The teaching of anthropology has long focused on the study of the differences between human groups, and neglected the analysis of the variability inherent to all populations. It oriented itself to a typological thinking that tends naturally to highlight the variability between populations, while minimising intrapopulational variability (i.e., variability within a population).

Being different is inherent in our human (cf. animal) nature and not a sign of inequality. The major differences are generally within populations and not between them. We are 6 billion people, all different and enriched by our differences. Why would homogeneity bring us more happiness? Humanity supports these differences and tolerance to difference should remain our leitmotiv: a tolerance that is not only cultural and philosophical, but which also implies respect for our biological differences.

[*] E-mail: scharles@vub.ac.be

HISTORICAL INTRODUCTION

For a very long time indeed, descriptions of foreign populations were regularly filled with profound prejudices. This attitude ("autrisme" or “separatism” or “apartheid”) implies that "the other" is systematically considered as being less perfect, included (and foremost) in psychological and sociological terms, what permits, in fact, to justify the existing discrimination. Throughout history, war and colonisation is thus "scientifically justified".

The teaching of anthropology has long focused on the study of differences between human groups and neglected analysis of the variability inherent in all populations (Susanne, 2003). Anthropology has engendered a typological thinking that tended to naturally highlight the variability between populations, while minimising the intrapopulational variability.

The notion of "race" is engrained in the general large public (the notion of "great race" based on the skin colour), and our task as educators is to show that “race” has no biological basis and that the study of human biology does not indicate clear subdivisions in the human species, contrary to what some might imagine.

The concept of "race" also includes a geographical component, with different "races" formed by geographical isolation in varied and particular living and environmental conditions. However, these large geographical groups form populations that have never been isolated at reproductive level and that have never been biologically (or genetically) pure (Templeton, 1998).

To characterise human populations according to their physical traits, such as skin pigmentation, hair colour and structure, etc., is to propose that these characteristics are influenced by genetic factors. However, characterisation of populations, not by their physical traits, but by gene frequency leads to contradictory results. On one hand, as, genetically speaking, the human species is very uniform compared to other animal species, the biggest genetic variations are observed between individuals of the same population. On the other hand, because genetic variations correspond to gradients of variation of gene frequency (clines), there is no gene that is typically Caucasian or Mongoloid or Basque.

It was only after the Second World War that human diversity came to be regarded as the result of the micro-evolutionary process, whereby gene pools of populations are differentiated under the effect of natural selection and genetic drift. Thus, races imagined as immutable came to be seen as fleeting clusters of alleles. These populations were constantly mingling, adapting to local conditions, joining or separating. With the advent of molecular biology in the 1960s, phenotypes were abandoned in favour of genetic data as a basis for studying human biological makeup. In 1950, a *Cold Spring Harbour symposium* of quantitative biology, called *Origin and Evolution of Man*, concluded that racial classification no longer has any scientific meaning and that researchers should show more interest in the evolutionary process and the genesis of diversity within the human species. In the same spirit, came that several UNESCO declarations: the *Statement on Race* of 1950, *Statement on the Nature of Race and Race Differences* of 1951, *Proposals on the Biological Aspects of Race* of 1964 and the *Statement on Race and Racial Prejudice* of 1967. These texts partially address anthropological preoccupations, but also provide answers to political and ethical questions and an attempt to avoid the use of biological data erroneously to uphold various forms of racism.

Today, the problem of "race" may seem outdated. In fact, our students are well aware that humans no longer consider themselves as the centre of the world. The universe has become the centre to be protected from those same humans. The biosphere is intrinsic valued, above the value of the species *Homo sapiens*, itself. Thus, a crisis is created as the limited capacities of the biosphere force us to find a new equilibrium between Man and Nature. Human evolution is a story of symbiosis, of control and much later of the domestication of Nature. It is also a story of human violation of Nature exacerbated by the unrestrained development of technology and population density.

Lagging behind, we are still preoccupied with Man as an island outside Nature; we might even have the tendency to consider humans as being above Nature because people develop original qualities and may seem to conquer Nature. Humans are, however, only animals who followed a process of humanisation. So, let us accept that Man is not opposed to Nature, but is part of it and is dependent on it.

STEREOTYPES PERSIST

In typological and racial classification, it is hypothesised that races in their pure state existed in the past, before migration caused large-scale mixing. With this reasoning, one forgets that migrations have always existed and thus gene flow has as well. As they say: "when groups meet they may or may not bleed, but they always breed".

In view of the evidence of continuous distribution (as in *clines*) of morphological or biochemical traits, which renders each "racial" demarcation arbitrary, anthropologists today prefer to reject the concept of typology. However, in current language, the term "race" is still used to define social groups in a complex society, without any biological significance whatsoever. Hindus are considered as "white" in the USA but as "coloured" in the UK (as are Chinese, Pakistani and Malaysians). In the South Africa of apartheid, a separate race was opined (whereas the "coloured" race of South Africa represents people of mixed origin). A first generation of mixed heritage (mulatto) of a parent of African origin and a parent of European origin, who receives the same genetic information from his father as from his mother and who has thus an ancestral origin both African and European, is considered in virtually all societies as belonging to the black population. The biological reality thus does not represent the sociological situation.

Many anthropologists and geneticists, without neglecting the existence of differences between individuals of the human species, propose no longer to use the term race. Their intent is to replace it with other terms, such as population, group of geographical origin or ethnic group, where traits of cultural and social origin are taken into consideration. However, it is difficult to separate biology and culture in our species. Ethnicity is a complex mixture of biology, history, orientation and cultural practices, as well as language, religion and lifestyle.

Moreover, the term ethnicity stirs up as many problems as the term race. Bernis writes, "The ethnicity or the ethnic identity usually goes together with the evaluation of other societies. When one's own is evaluated as being superior we speak about ethnocentrism. Along history and in all societies multiple examples exist, from the use of ethnic politics and ethnocentrism which led to incorrect political situations, wars and genocides. The ethnical conflicts have often been manipulated and exacerbated to justify strategic, political or economic interests which are in fact the engine of conflicts. Good examples are what recently happened in Europe with the wars destroying the Yugoslav Republic, or the conflict of the Great Lakes in Africa or the Israeli-Palestinian conflict in the Middle East"(Bernis 2004: 214).

THE HUMAN DIVERSITY

Homo sapiens is a species characterised not only by its success in numbers but also by its dispersion all over the planet, from extremely cold climates to hot and humid zones, from coastal zones to high altitudes. In this way, *Homo sapiens* displays a real diversity of phenotypes.

It seems clear that the more than 6 billion human beings who inhabit this planet are different from one another. This is evident when, for instance, one needs to find a donor for transplantation of an organ. However, diversity is a relative concept that depends not only on considered traits (monofactorial, such as blood groups determined by a single gene, or multifactorial, as the height determined by numerous genes but also by mesological factors), but also on the level of observation (i.e., whether between populations or within populations).

The progress of the last decades in molecular technologies, such as recombinant DNA technology, gene cloning, the use of enzymes and the restriction of enzymes or endonucleases, permit analysis of other polymorphisms, such as RFLP (a polymorphism of the length of DNA fragments generated by the digestion of restriction enzymes) and nucleic acid sequencing. Researchers also use these new techniques to study differences between human populations. We are thus confronted with new methods, but with the same old problems as with classical morphological traits: a mixture of genetics, "race", culture, and ethnicity.

Ongoing human genome studies have revealed that genes that are specific to a population do not exist and that a genetic uniformity within the human species exists throughout its large geographical dispersion. These studies also show that the greatest genetic distances are observed between individuals and not between populations: about 85% of the total genetic variation exists between individuals of the same population, with about 7% between "great race" populations (i.e. French, Italians, Spanish), and also only 7% between the major human "races" (such as the African, Asian, and European groups) (Lewontin, 1991). Similar observations exist also for mitochondrial DNA distribution, where 94% of the genetic variation within a population (Melnick *et al.*, 1992).

Human beings, like other species, are a result of their evolution: this is a circumstance accepted by the scientific community. Different religions, however, have for a long time defended the idea of a creation and denied the significance of human fossils, while some creationist movements (such as the Intelligent Design) still promote faith in this domain (Susanne, 2005). However, scientific observations are clear, from discovery of a fossil human branch about 7 million years old to analysis in terms of comparative anatomy, embryological development, and more recently, in terms of biochemical and DNA data.

The mechanisms of evolution are general, applicable to all living species and not specific only to the human species. The evolutionary history of the human branch of the tree of life, including in modern human variation, is the fruit of factors such as natural selection, migration and genetic drift. The origin of variations between populations is due to several factors: genetic flux, with its homogenising effect, as well as selection and genetic derivation. Selection will favour different genes in different environments (typical examples are sickle-cell anaemia or lactose intolerance). Genetic derivation is at the root of random differences between populations, which create fast divergences in gene frequency. Actual human variability may be considered as the result of this biological evolution, and of a differential adaptation to distinct environmental situations. Evolutionary forces have modelled the human species, and human variability may be considered as a naturalistic concept. At the same time, human variability has a geographical component in that fact variability will be linked to geographical isolation of populations in diverse environments (Deligne et al, 2001).

Actual genetic variations reflect the history of our migrations and development of human populations of *Homo sapiens* since our African origin. In fact, paleontological data show two "out of Africa" periods, one around 1.5 million years ago, with fossils equivalent to *Homo erectus*; and the other about 200,000 years ago, with the first African *Homo sapiens*. This timespan is, thus, too brief to account for great genetic variations: greater homogeneity is observed in the human species than in many other species, including our closest living "relative", the chimpanzee. Additional evidence provided by genetic analysis is the greater variability in the African gene pool than in those of other continents: genetic variation outside Africa represents only a marginal part of the African variation, which corresponds to the African origin of the first human groups to migrate during the second "Out of Africa" episode, about 200,000 years ago. Thus, one way or another, we can say that genetically speaking, all humans are Africans.

ECONOMIC RACISM? OR RELIGIOUS RACISM?

Is there a "happy ending" to the question "race"? There should be, but racial discrimination clearly persists. Racism and the social construction of "race" continue to exist (Susanne *et al.*, 2003b). This situation might be linked to certain economic or religious factors.

Economic Factors

The idea that the occidental world leads the boat of human history towards the secure port of liberal economics is in immediate contradiction in the light of environmental destruction. The reduction of reserves of various natural resources is also linked to the exponential growth of the human population. *Homo sapiens* has been replaced by the *Homo economicus (universalis)* (Crutwell, 1995). Today, the world has become a global village where only profits matter.

The negative sides of this economic "wonderland" are, however, present. The differences between North and South continue to grow. But it is hard to imagine that the actual politics of globalisation, coupled with a growing tendency toward social exclusion, violence, degradation of the environment, and cultural and political underdevelopment can continue without end. This situation is not only immoral but also inhumane. Populations that are forced to live in harsh conditions, knowing through the media what exists in the rest of the world, will not remain quite forever, without protesting.

Demographic Factors and Migrations

The demographic explosion also has numerous consequences for the degradation of the environment and the quality of life. At an ethical level, our societies, and the Church in particular, should understand that procreation without consideration for quality of life and respect for life is unacceptable. How else can we deal with the great favellas of South-American megalopoli for example, or the situation of abandoned children, drug slaves, prostitutes or even *"res corporalis"* of organ dealers.

Drastic efforts to reduce the population should be undertaken to ensure an acceptable quality of life and a sustainable level of development for future generations (Brundland, 1987). Solutions are possible as long as everyone participates—citizens by reducing their consumption and even their quality of life as individually, scientists by proposing more effective ecological techniques, industry by investing more in production that respects the environment, and politicians by using tools (laws, norms, taxes, ...) that favour a "greener" basis for sustaining our ways of life. Sustainable development can take place within these conditions, but only with the participation of the whole of society. This programme is, however, far from an enticing electoral platform.

Another consequence of demographic pressure is migratory pressure. Migrations have, of course, taken place all through our prehistory and our history, and have created both diversity and interbreeding. Are the migrations of today really more complex? Haven't migrations always had economic, ecologic and political origins (wars and conflicts)? But our world has become a "global village" where education and televisual information diffuse and idealise images relating to "rich" societies, and these images stimulate migrations of those who no longer wish to live in conditions of poverty as they seek a better life in an unknown place. Although education is improving in these poor regions, unemployment is so prevalent that not only migration but also the flight of brainpower is the result. In the end, it is not the poorest people who migrate, but those who have resources with which to make an exit. Thus migratory flux is not a solution for those countries—it is a drain but not a plug in a failing situation.

Migration also provokes the growth of racism, as well as the emergence of neo-fascism or extreme right views, which use hostility towards migrants to attract followers. Moreover, we can see that this racism has a significant impact in various European countries (Solomos & Wrench, 1993). Even if differences between its nations exist, trends throughout Europe are similar: transformation of the social infrastructure, economic stagnation, contraction of social advantages, "racial" problems in the workplace, housing and education, as well as continued isolation and ghettoisation of migrants of second and third generations (Heisler, 1991).

The ancient concept that "they are not like us" often becomes rewritten as "they do not want to be like us". In very subtle ways, political right movements reformulate and use the concept of "the right to be different" that humanist movements use to defend the rights of certain populations that are the object of prejudice and racism. These political right movements even transform themselves into defenders of ethnic pluralism to justify discrimination and expulsion of immigrants and their families.

By using migration as an issue on their political agenda, new forms of racism are insinuated and receive democratic support in the votes of the extreme right, which paradoxically introduced the germs of the denying of democracy. Racism "ethnicises" exclusion, inferiority, marginalisation and social hierarchy (Susanne *et al.*, 2004).

In the past, races were supposed to protect their biological purity, today, each culture is supposed to preserve its identity.

Religious Factors

Let us be critical about all forms of philosophies of exclusion, nationalism and religious fanaticism that stimulate fear, aggression and hatred. For all theocratic regimes, emotions are more important that intelligence because these regimes seek adhesion without discussion or reflection on the values of other sects. Rationality and critical thinking, which allow distancing from manipulation, are the enemy of fanaticism. Thus, we owe it to ourselves to continue the eternal battle for reasoned choice.

Democratic societies encourage development of tolerance as a desirable attribute, but at the same time we must be mindful that tolerance can actually have its limits. Tolerance must be bilateral and not influenced by fundamentalist minorities. Fundamentalist groups exist in all monotheistic religions, although they do not represent all practitioners of these religions. It goes without saying that the sometimes intolerable positions of fundamentalist groups cannot be imposed on the entire society. Fundamentalism represents a manifestation of a political venture trying to impose on society, from the individual to the State, rigorous values of religion, while instrumentalising God to create commerce and influence politics. This group cannot accept the concepts of pluralism of ideas or diversity of life. Freedom of expression is a nightmare for extremists of the three major monotheistic confessions, Christian, Jewish and Islam. These extremist movements even talk about racism (that is, antireligious racism) every time religion is criticised, their strategy being clearly to limit freedom of expression. Humanism must fight racism, but also violence, sexism and intolerance, even if it is of religious origin. Is it not intolerance when religious or community pressures try to curtail the teaching of evolution (Susanne, 2005)? What about proposals in the UK to forbid teaching about the holocaust because it would offend some extreme Muslim groups, who claim that holocaust did not exist?

Our societies need to remain secularised and avoid participation in a reign of racism. To promote rationalism and oppose fundamentalism is not an antireligious racism. Tolerance is not intolerance.

CONCLUSION

"Racial" classifications are the reflection of social concepts and do not have a biological basis. Contrary to popular interpretation of the existence of "great races", based on traits perceived as essential (i.e. the colour of the skin), this notion has no biological validity and its differences are negligible compared to the genetic differences existing between individuals of the same population. To divide human populations into a certain number of groups provokes arbitrary subdivisions and is therefore not natural. In fact, populations constitute local bio-cultural units that undergo perpetual change. The human species forms a continuum (or clines) following geographical gradients and ecological barriers: genetic flows are constant and differences between populations do not imply judgments of values.

The notion of "race" has actually lost its scientific value, even if it preserves a signification of social identification. The appropriate position relating to racism is not to deny the differences that exist between populations, or to deny the need of humans to identify themselves to a group, but to ensure that different groups have the same access to resources and that no group, nor an individual of that group, would be discriminated against.

The biggest differences and real barriers between human populations are of cultural origin (linguistic, behavioural, clothing differences, etc.). Nonetheless, let us not imagine that knowing that "race" is a myth might eliminate racism and its discrimination automatically. Racism remains a reality. The myth of the inequality of races is a political vision, put back on the agenda by the ideology of the extreme right and its ethnocentric nationalisms.

The 19th and 20th centuries were accompanied by profound conceptual changes: in the 19th century, the progress of biology undermined anthropocentrism and in the 20th century undermined the progress of social anthropology ethnocentrism (Susanne & Hens, 1998). "Is Man unique?" is a simple question that is central to biological anthropology and at the origin of many false reasoning. The anthropologist needs to reply that Man is indeed unique, as all other animal species are unique, and that a chimpanzee is not a gorilla and a cat not a lion. In the meantime, the anthropologist should relativise this unique character: the animal world responds to the same biochemical principles. Uniqueness is naturally the case with all mammals and primates. We share a genetic similitude of more than 99.5% with the chimpanzee.

Knowledge of the genome is part of daily life in relation to the institutions, its tribunals, the insurance companies, ethnic identification, and the medical possibilities, etc. that it engenders. All results of genome study and application show that populations do not differ by the presence of specific genes, and that they are characterised by the same genes but in the different frequencies. The spectacular progress in human genetics confirms the observations of a genetic variation inherent to each population. The fundamental contribution of the studies in human genetics is not to classify populations, but to establish affinities between populations by reconstituting the history of populations, their ecosystems, their migrations and their fusions. In other words, genetics aims to understand the history of "racial" integrations, which are, in the end, the history of humanity (Bernis, 2004).

Being different is our human (and animal) nature. Being different is not a sign of inequality, as the major differences are within populations and not between them. The Earth's population constitutes some 6 billion people, all different and all enriched by these differences. Why would homogeneity bring us more happiness? Humanity engenders differences, and tolerance of differences should remain our leitmotiv: tolerance is not only a cultural and philosophical stance: it also must imply respect for our biological differences.

Man often fears his freedom: it is a tool we cannot completely master. We often prefer stereotypes and even totalitarian ideological remedies to calm our feelings of inferiority and our metaphysical anxieties. Man needs to be aware of his own animal nature to be able to take charge of "his own animal".

Let us remain prudent: in the actual context of "everyone for themselves", ethnic exclusion and racism gain way again.

"Fear is ignorance," said Ghandi. Let us try to better understand our differences, that they may seem minimal and worthy of study, but not of dispute and conflict. Manipulation is not only genetic, but is embodied in words.

Let us remain careful of all manipulative speeches and guard against the return of the inequality of "races". Let us continue the free speech of education as the only tool for teaching our generations to avoid discrimination and to remain upright and free.

REFERENCES

Bernis, C. (2004). Ethnicity and racism. In C. Susanne (Ed.), *Societal Responsibilities in Life Sciences (Journal of Human Ecology*, Special Issue 12, pp. 213-221). New Delhi: Kamla-Raj Enterprises.

Brundlandt, G.H. (1987). *Our Common Future*. Oxford: Oxford University Press.

Cruttwell, P. (1995). *History out of control.* Devon, UK: Green Books Ltd.

Deligne, J., Rebato, E., & Susanne, C. (2001). Races et racisme. *Journal des Anthropologues* 84: 217-235.

Heisler, B. (1991). A comparative perspective on the underclass: questions of urban poverty, race and citizenship? *Theory and Society* 20: 455-483.

Lewontin, R.C. (1991). *The Doctrine of DNA.* London: Penguin.

Melnick, D.J., Hoelzer, G.A., & Honeycutt, R.L. (1992). Mitochondrial DNA: its uses in anthropological research. In E.J. Devor (Ed.), *Molecular Applications in Biological Anthropology* (pp. 176-223). Cambridge: Cambridge University Press.

Solomos, J., & Wrench, J. (1993). Race and racism in contemporary Europe. In J. Wrench, & J. Solomos (Eds.), *Racism and migration in Western Europe* (pp. 3-16). Oxford: Berghan.

Susanne, C. (2003) .L'histoire de l'évolution humaine. In C. Susanne, E. Rebato, & B. Chiarelli (Eds.), *Anthropologie Biologique: Evolution et Biologie Humaine* (pp. 19-30). Bruxelles: De Boeck Université.

Susanne, C. (2005). Teaching of (human) evolution in danger? *Studia Bioetica*, 2, http://utopia.duth.gr/~xirot/BIOETHICS/.

Susanne, C. & Hens, L. (1998). From anthrocentrism to ecocentrism. In C. Susanne, & L. Galle (Eds.), *Ecotechnie and Sustainable Development* (pp. 35-56). Szeged: Szeged University Press.

Susanne, C., & Rebato, E. (2004). Race, a stereotype. In C. Susanne (Ed.), *Societal Responsibilities in Life Sciences (Journal of Human Ecology*, Special Issue 12, pp. 223-226). New Delhi: Kamla-Raj Enterprises.

Susanne, C., Rebato, E., & Deligne, J. (2003). Races et racisme. In C. Susanne, E. Rebato, & B. Chiarelli (Eds.) *Anthropologie Biologique: Evolution et Biologie Humaine* (pp. 643-652). Bruxelles: De Boeck Université.

Templeton, A.R. (1998). Human races: a genetic and evolutionary perspective. *American Anthropologist* 100: 632-650.

INDEX

B

C

F

G

H

I

J

K

L

M

N

O

P

Q

R

S

T

U

V

W

Y